Preservation of Cultural Heritage and Resources Threatened by Climate Change

Special Issue Editor

Chiara Bertolin

MDPI • Basel • Beijing • Wuhan • Barcelona • Belgrade

MDPI

Special Issue Editor
Chiara Bertolin
Norwegian University of Science
and Technology
Norway

Editorial Office
MDPI
St. Alban-Anlage 66
4052 Basel, Switzerland

This is a reprint of articles from the Special Issue published online in the open access journal *Geosciences* (ISSN 2076-3263) from 2018 to 2019 (available at: https://www.mdpi.com/journal/geosciences/special_issues/Preservation_Cultural_Heritage_Climate_Change)

For citation purposes, cite each article independently as indicated on the article page online and as indicated below:

LastName, A.A.; LastName, B.B.; LastName, C.C. Article Title. *Journal Name* **Year**, *Article Number, Page Range.*

ISBN 978-3-03921-124-1 (Pbk)
ISBN 978-3-03921-125-8 (PDF)

Cover image courtesy of Alessandra Bonazza.

Preservation of Cultural Heritage and Resources Threatened by Climate Change

Contents

About the Special Issue Editor

Chiara Bertolin obtained her Master's degree and Ph.D. in Astronomy at the Padua University, Italy. Since 2006, she has been a researcher at the National Research Council – Institute of Atmospheric Sciences and Climate in Padua, working on several EU funded projects (i.e., Millennium, Sensorgan and Climate for Culture) and international scientific cooperation. Over the years, she has developed cross-disciplinary research interests on climatology, microclimate, and environmental assessment for cultural heritage preservation. To date, she has authored 40 papers in international journals, one monograph, 10 book chapters and made more than 60 contributions to international conferences proceedings. She also holds one patent. Since 2013, she has been a member of the European Committee for Standardization. She has worked as a scientific advisor for UNESCO; the Venice Civic Museums Foundation; the Diocesan Museum in Udine, Italy; and the Norwegian Institute for Cultural Heritage Research in Oslo, Norway. Since 2016, she has been an Associate Professor at the Norwegian University of Science and Technology (NTNU) in Trondheim, Norway within the Onsager Fellowship Programme, part of NTNU Research Excellence. Today, she is the coordinator and principal investigator of the Young Research Talents project: SyMBoL - Sustainable Management of heritage Buildings in a Long-term perspective, started in September 2018. SyMBol aims to solve the debate about the appropriate environmental conditions to preserve, in this time of climate change, the most precious heritage buildings in Norway; i.e., the Stave Churches and their distempered paintings.

Preface to "Preservation of Cultural Heritage and Resources Threatened by Climate Change"

This book presents a print version of the Special Issue of the journal Geosciences dedicated to the "Preservation of Cultural Heritage and Resources Threatened by Climate Change." With a wide spectrum of data, case studies, monitoring, experimental and numerical simulation techniques, the overall goal of this Special Issue was to provide the most current state-of-the-art research on the recognition, analysis, and management of natural and human-induced climate change impact on cultural heritage. In the 10 papers collected in this volume, readers will recognize the importance of a multidisciplinary approach (for example, involving materials, environmental and computer science knowledge) in identifying predominant risks for cultural heritage preservation in the time of climate change. Among the articles published in the Special Issue, three research studies are based on the exploitation of a broad range of data derived from preventive conservation monitoring. Two are further focused on climate data and numerical modeling data for assessing environmental impact and climate change effects. Four papers are focused on a well-assorted sample of decay phenomena occurring on heritage materials, e.g. surface recession and biomass accumulation on limestone, depositions of pollutant on marble, salt weathering on inorganic building materials, and the weathering process on mortars. Finally, the remaining paper is devoted to examining the perceptions of experts involved in the management of cultural heritage on adaptation to climate change risks. Thirty-five authors from three different continents (Central America, Europe, and Oceania) contributed to the Special Issue, showing results from local to regional-scale study areas in the Scandinavian Peninsula, United Kingdom, Belgium, France, Italy, Greece and Panama, wide enough to attract the interest of an international audience of readers. The articles collected here will hopefully provide different, useful insights into advancements in emerging technologies for the monitoring and the future forecasts of key degradation phenomena in both organic (e.g., wood) and inorganic (e.g., marble, cement mortar) heritage material. Finally, these research studies confidently highlight new ideas, approaches, and innovations in the analysis of various types of decays (e.g., surface recession, biomass accumulation, salts weathering) in a number of sensitive environments (e.g., indoor museum environments, urban, coastal, and rural).

Chiara Bertolin
Special Issue Editor

geosciences

MDPI

Editorial

Preservation of Cultural Heritage and Resources Threatened by Climate Change

Chiara Bertolin

Department of Architecture and Technology, Norwegian University of Science and Technology, Alfred Getz vei 3, 7491 Trondheim, Norway; chiara.bertolin@ntnu.no

Received: 24 May 2019; Accepted: 27 May 2019; Published: 3 June 2019

Abstract: With a wide spectrum of data, case studies, monitoring, and experimental and numerical simulation techniques, the multidisciplinary approach of material, environmental, and computer science applied to the conservation of cultural heritage offers several opportunities for the heritage science and conservation community to map and monitor the state of the art of the knowledge referring to natural and human-induced climate change impacts on cultural heritage—mainly constituted by the built environment—in Europe and Latin America. The special issue "Preservation of Cultural Heritage and Resources Threatened by Climate Change" of *Geosciences*—launched to take stock of the existing but still fragmentary knowledge on this challenge, and to enable the community to respond to the implementation of the Paris agreement—includes 10 research articles. These papers exploit a broad range of data derived from preventive conservation monitoring conducted indoors in museums, churches, historical buildings, or outdoors in archeological sites and city centers. Case studies presented in the papers focus on a well-assorted sample of decay phenomena occurring on heritage materials—e.g., surface recession and biomass accumulation on limestone, depositions of pollutant on marble, salt weathering on inorganic building materials, and weathering processes on mortars in many local- to regional-scale study areas in the Scandinavian Peninsula, the United Kingdom, Belgium, France, Italy, Greece, and Panama. Besides monitoring, the methodological approaches that are showcased include, but are not limited to, original material characterization, decay product characterization, and climate and numerical modelling on material components for assessing environmental impact and climate change effects.

Keywords: cultural heritage; climate change; decay; preventive conservation; mitigation actions; heritage materials; indoor climate; outdoor climate

---

1. Introduction

Changes in preservation conditions due to climate-related decay processes are unavoidable phenomena for both movable and immovable cultural heritage (CH). The knowledge of the mechanisms governing these processes and their real effect on changing heritage significance will allow the rational use of heritage materials, as well as the anticipation of their behavior beforehand, in order to succeed in preventive conservation, heritage management, and eventual restoration.

The degradation progress, which depends on external agents of decay, exposure, the intrinsic properties of the material to be studied, and object construction vulnerability, is nowadays exacerbated by both anthropic factors and the impact of climate change (CC).

Due to more frequent and severe weather events, greater exposure, ageing of materials, and the existence of previous conservative interventions, the need for adapting cultural heritage to anthropic and climate change-related effects is becoming more and more urgent. Within cultural heritage, the risk from climate change is more pronounced for the built environment, where the right adaptation interventions should be chosen properly considering the buildings' capacity to change due to its protection status under the law, the principle of preservation, and the need to apply effective mitigation actions. The International Council on Monuments and Sites (ICOMOS) principles of preservation [1] ask to

-   understand and respect CH and its significance;
-   be cautious in designing interventions;
-   respect authenticity and integrity;
-   propose reversible interventions to–as much as possible–keep intact existing original materials;
-   prioritize preventive and effective care;
-   prioritize minimum intervention: "do as much as necessary and as little as possible";
-   propose compatible design solutions, i.e., use adequate materials, techniques, and detailing with regard to material and physical–chemical–mechanical interactions between the new and the existing;
-   enhance the use of cultural assets and regularly programmed maintenance necessary to extend the service life of the CH;
-   enhance multi-disciplinary action—i.e., call upon skill and experience from a range of relevant disciplines.

For movable objects (often preserved indoors in museums or historical buildings), climate change-induced risks manifest themselves in greater difficulty of management, due to higher costs for cooling/heating demand related to maintaining appropriate environmental conditions, as requested by the American Society of Heating, Refrigerating and Air-Conditioning Engineers (ASHRAE) and the European Committee for Standardization (CEN) standards, as well as higher costs for conducting preventive conservation due to the new appearance or acceleration of decay phenomena.

The Council of Europe's European Heritage Strategy for the 21st Century [2] calls for more reliable quantified information on the impact of climate change on cultural heritage, as the changing climate, speeding up the rate of degradation and the risk of loss of value, is affecting the organizations who take care of cultural heritage, resulting in difficulties in managing the maintenance of heritage buildings outdoors and the indoor environments caused by increasing costs and lack of funds. As a whole, this in turn affects the cultural tourism sector, local and regional economies, their traditional practices in maintenance and conservation, as well as their use of resources and adaptation planning options.

However, limited research has been accomplished to date on the process of preserving CH threatened by CC, as very few tools or methods exist to collect and analyze data on the actual situation, and at the same time estimate the ongoing and expected risks.

The main objectives of this special issue are

-   to make the point about the ongoing research in the field;
-   to present new data, methods, and techniques that can be used by a wide community of researchers and conservators to better understand degradation phenomena affecting heritage materials, and to assess the actual and expected impact of CC;
-   to provide guidance on conservation principles and standards to follow in order to enhance awareness on preventive conservation and long-term planned conservative interventions in the time of CC;
-   to develop mitigation and adaptation capacity throughout the wide range of stakeholders involved, as it is urgent to respond to CC now.

2. Overview of the Special Issue Contributions

The special issue (SI) of *Geosciences* titled "Preservation of Cultural Heritage and Resources Threatened by Climate Change" has been launched to take stock of the existing, but still fragmentary knowledge on this challenge, and to enable the heritage community to respond to the implementation of the Paris Climate agreement. At the European level, only two projects, "Noah's Ark" [3] and

"Climate for Culture" [4], have finalized their research on these issues. The SI encompass 10 open access papers presenting research studies based on the exploitation of a broad range of data deriving from preventive conservation monitoring [5–7] and climate or numerical modelling on material components for assessing environmental impact and climate change effects [8,9]. These papers are focused on a well-assorted sample of decay phenomena occurring on heritage materials, e.g., surface recession and biomass accumulation on limestone [10], depositions of pollutant on marble [11], salt weathering on inorganic building materials [12], and the weathering process on mortars [13]. Finally, one paper [14] is devoted to examining the perceptions of experts involved in the management of cultural heritage on adaptation to climate change risks.

Table 1 summarizes the distribution of data and techniques used in each paper, the target decay type monitored or simulated on one or more heritage materials, and the considered climate change scenario and assessment time.

Table 1. Overview of data, techniques, target decay, heritage material, and climate change scenario, with the assessment times presented in the 10 open access research papers composing the special issue (SI) "Preservation of Cultural Heritage and Resources Threatened by Climate Change" of *Geosciences*. The papers are in order of publication.

Paper Reference and DOI	Data and Processing/Analyzing Method	Type of Decay and CH Material	CC Scenario and Assessment Time
Anaf et al. [5] doi:10.3390/geosciences8080276	Monitoring data; Indoor Air Quality (IAQ) index for heritage application	Overall IAQ risk; air-mixed materials/objects	Real time or post-adaptation measures assessment
Ciantelli et al. [10] doi:10.3390/geosciences8080296	Meteo-climate data (Temperature, Relative Humidity, rain) and main construction materials characterization	Pollution and salt crystallization cycles; masonries	Reference period: 1979–2008 and future: 2039–2068; EC: Earth global climate model with high GreenHouse Gas (GHG) emissions
Sesana et al. [14] doi:10.3390/geosciences8080305	Stakeholder interviews and participatory workshop; best practices in adaptation	Hazards exacerbated by CC (flood, landslides, Sea Level Rise)	Reanalysis of past experience for future strategic planning and preparedness
Carroll and Aarrevaara [6] doi:10.3390/geosciences8090322	Materials and structures classification; urgency index	T and precipitation trends resulting from CC; Mixed materials on buildings	Climate trends and assessment of past adaptation measures
Dotsika et al. [13] doi:10.3390/geosciences8090339	Isotopic data from mortar samples collected at different depths; stable isotope analysis	Chemical decay due to pollution and environmental conditions, and the secondary decay mechanism of carbonate formation from salt weathering and biological attack; mortars	Analysis of weathering progression from Hellenistic, Late Roman, and Byzantine mortar layers
Loli and Bertolin [8] doi:10.3390/geosciences8090347	Climate for culture maps and building protection levels by law; decay level estimation and allowable interventions	Chemical, biological, mechanical; mixed buildings materials	Far future: 2071–2100; Regional Model, rapid economic growth (A1B) and representative concentration pathway to 4.5 W/m^2 radiative forcing value within 2100 (RCP4.5) scenarios
Fermo et al. [11] doi:10.3390/geosciences8090349	Monitored data on deposited aerosol particulate matter (PM) on quartz filters and main ions, atmospheric pollution data; chemical characterization	Environmental deposition; marble and surrogate substrates	Present: 2014–2017
Haugen et al. [7] doi:10.3390/geosciences8100370	Environmental monitoring (T, RH) and moisture content on wood; zero-level registration and interval-based registration system for relevant indicators	Mixed decays on stone and wooden historical buildings	Present-day monitoring is planned to continue for 30–50 years
Bylund Melin et al. [9] doi:10.3390/geosciences8100378	Monitoring of T and RH to simulate moisture content in wood; monitoring, experimental tests, and simulations using WUFI Pro software and simplified mathematical models	Moisture diffusion and transport that can induce primary mechanical decay and secondary effects; wood or hygroscopic materials	Building simulations and climate change scenarios
Menéndez [12] doi:10.3390/geosciences8110401	Phase change phenomena of common salts; comparison of predicted changes in weathering driven by single or mixed salts.	Salt weathering; historical buildings	Present and future simulation of weathering

Data, Methods, and Decay on Cultural Heritage Material under Climate Change Scenarios

In their paper, Anaf et al. [5] present a new tool to help heritage guardians in processing and evaluating monitored data in museums or heritage buildings. The work explains the backbone of their proposed Indoor Air Quality (IAQ)-calculating algorithm, from the recognition of deterioration agents, to conversion functions for calculating the level of risk in specific materials or objects, to the way in which weight is attributed to key risk indicators (KRIs). The combination of all KRIs constitutes the overall IAQ index. The authors clearly present this new tool, visually processing sets of multi-monitored data applied to canvas painting, restrained wood, and copper preserved in a church in Belgium. This allows the identification and discussion in detail of the potential and limitations of the IAQ index for assessing the effectiveness of mitigation actions implemented by a heritage institution. Finally, this tool helps to better manage the indoor environment, in order to adapt it to a changing climate.

Keeping the focus on data assessment derived from environmental monitoring in museums as heritage institutions, Carroll and Aarrevaara [6] reanalyze literature findings about the range of local weather- and climate-related factors that contribute to the degradation of cultural heritage buildings and structures over time. These factors manifest themselves by speeding up the rate of degradation. The authors propose a method to collect information about where best to concentrate cultural heritage site preservation resources in the future, based on an urgency rate index. This proposed numerical scale index ranging from 1 to 10 can be applied to several CC categories (i.e., warmer climate, longer growing season, increased precipitation, severe rain, and extreme winds), to be evaluated both in terms of the rate of change (i.e., the increase in °C/yr, days/yr, mm/yr, mm/hr, or m/s, respectively) and a visual inspection on the affected structures or materials. Arctic regions are predicted to face the greatest increase of warming in winter time. The temperature is expected to rise by 3–4 °C by 2050. Rainfall in Nordic countries will increase by about 10% on an annual level. Exceptionally, the west coasts of Norway and Finland might face as much as a 20–30% increase in rainfall in winter periods. This will turn to cause favorable conditions for fungal growth and pest damage. To highlight such risks, and to show the potentiality of the proposed model, Carroll and Aarrevaara apply it to evaluate the outdoor conditions of a Finnish farmhouse complex consisting of several buildings in Finland. The method, as well as the discussion of its results, serves the purpose of prioritizing cultural heritage materials and elements for protection against the ravages of climate change, therefore helping conservators and heritage managers in planning for adaptation or mitigation steps.

From the majority of the works presented in the SI, it is clear that a long-term strategy to adapt to climate change must be based on risk assessment, adaptation measures, and monitoring. Specifically, due to the still-high degree of uncertainty in simulations, long-term monitoring of the actual impact of climate change is necessary to better understand the effects of climate change on historic buildings. Monitoring can be used to observe and analyze decay progress, as already stated in [5,6], in order to provide reference data to improve the results of simulation models; inform decisions on adaptative or corrective actions; raise awareness among property owners, heritage managers, and citizens; and to gain political and economic support locally, regionally, and nationally. Looking at systematic, well-planned, long-term management, Haugen et al. [7] presents an innovative methodological approach that is able to record climate change-induced decay on historic buildings and interiors at the national scale. The additional challenge here is the long time frame needed to discern the climate change signal from the natural variability of the climate. This time frame (>30 years) exceeds the scope of most research funding schemes, and there are also practical difficulties in maintaining such long projects with regard to administration, staff continuity, data retrieval and storage, etc.

In the first part of their paper, Haugen et al. [7] present the generic framework of the novel methodology. This framework is based on a review of existing approaches to climate change monitoring of cultural heritage, as well as the experiences from the Norwegian pilot project known as "Methods for Monitoring the Effects and Consequences of Climate-Related Degradation of Buildings", which proposes, for the first time, zero-level registration and an interval-based registration system focused on relevant decay indicators to detect the effects of climate-induced degradation. The second part of the paper presents, as a case study, the implementation of a newly started, long-term monitoring campaign on 45 medieval buildings distributed over the entirety of Norway. Thirty-five of these buildings are dated to before 1537, and include wooden buildings as well as 10 medieval churches built in stone, while the remaining 10 buildings are situated in the World Heritage sites of Bryggen in Bergen, on the west coast of Norway, and in Røros. Last but not least, the importance of the early involvement of researchers and stakeholders (e.g., the directorate of cultural heritage, conservators, heritage institutions, staff and building owners, etc.) from an early stage is fundamental for the success of such a long-term monitoring program.

Moving from research studies based on the exploration of data from preventive conservation monitoring to climate and numerical modelling on material components for assessing climate change effects, Loli and Bertolin [8] present multi-risk scenarios of CC on building materials, using data from the European Union (EU)-funded project Climate for Culture (CfC) [4]. The authors employ a modified version of the risk assessment method developed in the CfC project to take into consideration the proper adaptation intervention to be applied on historic buildings. The authors link the majority of climate-induced decay variables, describing mechanical, chemical, and biological decay on several building materials (e.g., masonry, concrete, and wood) and structures, with the buildings' capacity to change due to their protection status. The merging of the decay results with the building protection level becomes an indicator of the right level and time for intervention for climate change adaptation. The proposed method was then tested on 38 locations in Scandinavian countries to estimate the influence of climate change on future interventions on historic buildings [8].

A risk assessment matrix of deterioration highlights that, over the far future (i.e., 2071–2100), the risk of chemical and biological decays (outdoors) will slightly increase, especially in the southern part of the Scandinavian peninsula, while the mechanical decay of building materials kept indoors will generally decrease. This, for example, will require high-priority interventions for small, heavyweight buildings located in the area near Göteborg and Malmö, in order to adapt measures that minimize the climate-induced decay expected over the far future.

With a similar approach to the use of climate, building, and material modelling, Bylund Melin et al. [9] propose a study to increase the knowledge of climate-induced damage to heritage objects, which is essential to monitor moisture transport in wood. In fact, hygroscopic materials, such as wood, will gain and release moisture during changes in relative humidity (RH) and temperature (T). These changes cause swelling and shrinkage, which may result in permanent damage. To propose simulation models that are able to predict how the influence of climate change will modify the wood moisture content, and consequently the risk of shrinkage and swelling, is of primary importance. Bylund Melin et al.'s approach is completely multi-disciplinary, as they compare experimental data acquired in the laboratory, such as monitored temperature (T) and relative humidity (RH) at different depths inside wooden samples subjected to fluctuating climate over time, with novel methods, i.e., the use of hygrothermal building simulation software WUFI Pro to simulate object components, as well as a simplified model to calculate the moisture content. The conclusion was that both methods can simulate moisture diffusion and transport in wooden object with sufficient accuracy. In addition, both methods for predicting climate change data show that the mean RH inside wood remains rather constant, but the RH minimum and maximum vary with the predicted scenario and the type of building used for the simulation.

With a specific look at surface recession and biomass accumulation on limestone as a long-term climate change effect, Ciantelli et al. [10] present a case study in Latin America, in Panamá Viejo (a 16th-century building) and at the Fortresses in Portobelo and San Lorenzo (17th- to 18th-centuries). The authors first analyzed the main construction materials at the site level (i.e., masonries and limestone), adopting several investigation techniques (e.g., stereomicroscope, polarized light microscopy, X-ray powder diffraction, environmental scanning electron microscopy, ion chromatography), and then they analyzed changes in rainfall, RH, and surface air T as key drivers of the deterioration of cultural heritage. They applied future model predictions (running EC-Earth Global Earth System Model at high horizontal resolution) of these variables in damage functions to study the different kinds of material decays that might occur in the future. In particular, all functions they considered indicate an increase in surface recession, biomass accumulation, and cycles of dissolution and crystallization of halite in the future (2039–2068) with respect the past (1979–2008), especially in the North Coast, as shown by the analysis performed at the San Lorenzo and Portobelo areas. Nevertheless, the Panamá Viejo zone also shows an increment of surface recession and biomass accumulation; while considering the salt cycles, growth is projected to decrease. This work represents an important contribution to better understanding the possible future impact of CC on the heritage sites of Central America, and to support their management, restoration, and preservation.

With similar objectives Fermo et al. [11] present the results of the field exposure activity conducted between 2014 and 2017 on the marble façade of the cathedral in Milan, Italy. The authors performed a complete chemical characterization in real exposure conditions, quantifying deposited aerosol particulate matter (PM) and main ions on quartz filters and marble substrates. Through their monitoring strategy, they were able to discriminate between the compositions of the deposits, mainly depending on the type of substrate used (e.g., stone—Candoglia marble—substrates and quartz fibre filters as surrogate substrates), exposed on two sites of the cathedral façade at different heights. On the quartz filters, the carbonaceous component of the deposits was also investigated, as well as the color change induced by soiling, by means of colorimetric measurements.

The paper by Dotsika et al. [13] analyze 63 samples of mortars collected from lime and hydraulic mortars affected by environmental degradation, obtained from Hellenistic, Late Roman, and Byzantine historic constructions located at Kavala, Drama, and Makrygialos in northern Greece. The analysis of isotopic data allowed the re-creation of an ideal Hellenistic and Byzantine mortar layer to study weathering gradients. In fact, authors collected the first sample from the external layer, while the internal samples each were from 1 cm deeper than the previous, in order to monitor the moisture ingress. The obtained results indicate that a stable isotope analysis is an excellent tool to fingerprint the origin of carbonate, the environmental setting conditions of mortar, and the origin of CO_2 and water during calcite formation, as well as to determine the weathering depth and potential secondary degradation mechanisms, such as the recrystallization of calcite with pore water and salt attack.

Correspondingly, with a combination of methods described in [10,11,13], i.e., material analysis, damage function application, and climate change scenarios, Menéndez [12] estimates the salt weathering induced by climate change on built cultural heritage in 41 locations in France. In the analysis of phase-change phenomena, the author uses not only the two most common salts held responsible for decay, i.e., sodium chloride and sodium sulfate, but also others like calcium sulfate or mixtures of chlorides, sulfates, and nitrates of sodium, calcium, magnesium, and potassium. The novelty of this work is that it proposes a comparison between the predicted changes in salt weathering obtained from the presence of a single salt and a combination of different salts. The results achieved by Menéndez demonstrate how estimations of actual and future weathering depend on the selected salts. In addition, when using a combination of different salts, the weathering evolution is less favorable than when using a single salt.

Last but not least, Sesana et al. [14] examine, using semi-structured interviews, the perceptions of experts involved in the management of cultural heritage with regard to adaptation to climate change risks. This is a very sensitive topic, due to greater exposure to severe weather events; however, to date, limited research has been accomplished in the literature on the process of adaptation. In the paper, the authors report answers obtained by the contacted experts in the United Kingdom, Italy, and Norway, resulting from a participatory workshop organized with stakeholders on management methodologies that contemplate climate change impacts and examples of best practice. The work dispenses insights on opportunities and barriers in adaptation, including requirements for preparedness and future strategic plans for cultural heritage protection in the time of climate change.

3. Statistics, Bibliometrics, and Impact

The 10 research papers were published in the special issue between the end of July 2018 and early November 2018, with an average time of less than two months from first submission to online publication. Each manuscript was assessed via rigorous peer reviewing from two or more esteemed experts in the respective field.

The geographic distribution of the authors and research teams publishing in the SI and of the case studies are reported in Figure 1.

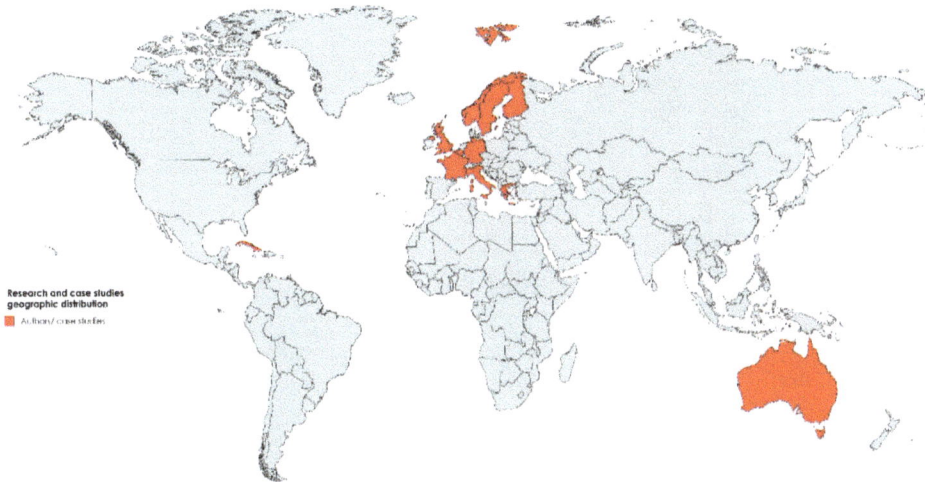

Figure 1. Geographic distribution of authors and research teams publishing in the special issue. Case studies are located in the same countries belonging to the authors, except for Australia, Cuba, and Germany, while a case study not highlighted in the map is located in Panama.

This is, of course, a sample of the whole scientific community working on climate change impacts on cultural heritage, although not an exhaustive representation. However, it already provides a glimpse of the widespread expertise of experimental research, field practice, building, and climate simulation, and proves how widely the conservation of cultural heritage is applied to investigate, mitigate, and adapt to the impacts of climate change.

Based on article metrics powered by PlumX on Scopus, overall, the published papers have already received eight citations in the indexed literature in the first few months after publication, with an average of almost one citation per paper, proving the immediate impact of the published research. Additionally, the item-level metrics provide insights into the ways people interact with the articles constituting the SI in the online environment. Beside the main scientific outcome of the citation metric, the other stronger interactions are in captures and social media areas, with an average of six captures as bookmarks, favorites, and reference manager saves by readers, and an average of six tweets, likes, and shares on social media. This could indicate that the SI papers are already being consumed and talked about.

Figure 2 highlights the disciplines and scientific domains on which the 35 authors of the papers published in the special issue are experts, as inferred from their history of publications from Scopus.

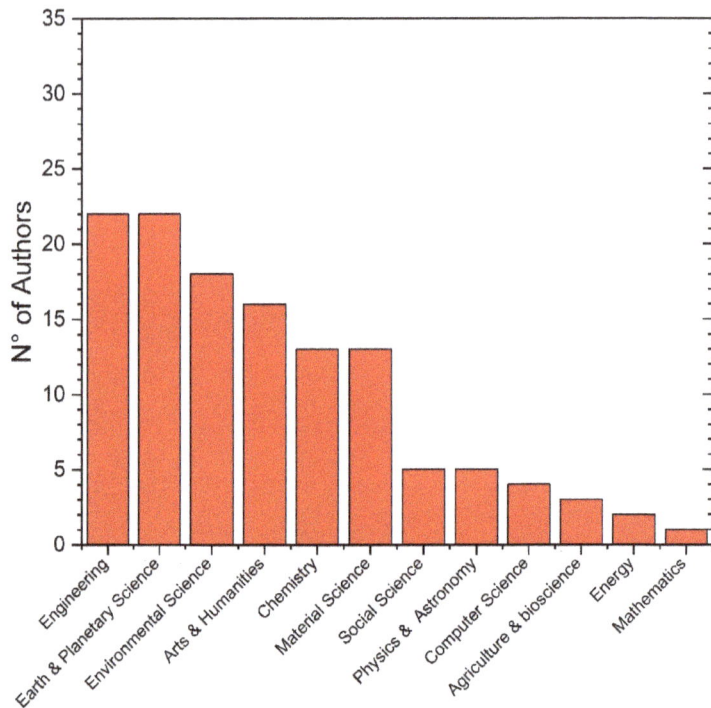

Figure 2. Expertise and scientific domain of authors publishing in this special issue, as inferred from their publications in Scopus.

Several situations can be observed generally:

(1) Within this SI, the common background among co-authors of the same paper and between authors of different works is on engineering, earth and planetary science, and environmental science. This make it clear the types of qualifications needed to understand the climate change at present and in the future.

(2) In each research team that has published a paper in this SI is present at least one or more experts on arts and humanities, chemistry, and material science, which are fields related to know-how in preventive conservation, the museum environment, and heritage and conservation science

(3) Authors with different professions have joined efforts to combine skills for building and climate simulation processing (e.g., physics and astronomy, as well as computer science, but again engineering and earth and planetary science) with arts and humanities or social science expertise.

Finally, using the data source for Scival metrics in the Scopus database, and specifically the Scival topic prominence, it is possible to know the topics in high-momentum areas. Prominence is in fact an indicator that shows current momentum by weighing three metrics for papers clustered in a topic: citation count, Scopus views, and average CiteScore. The topics treated within this SI (e.g., museums–buildings–preventive conservation; adaptation–climate change–vulnerability assessment) have an extremely high percentile prominence ranging from the 80 to 99.5 percentile, which is indicative as the authors are currently active in globally prominent topics. Figure 3 reports the visual result of the topic aggregation levels of this SI. It shows that the main core of expertise of the authors is on physics and astronomy (PHYS), chemistry (CHEM), chemical engineering (CENG), and materials science (MATE).

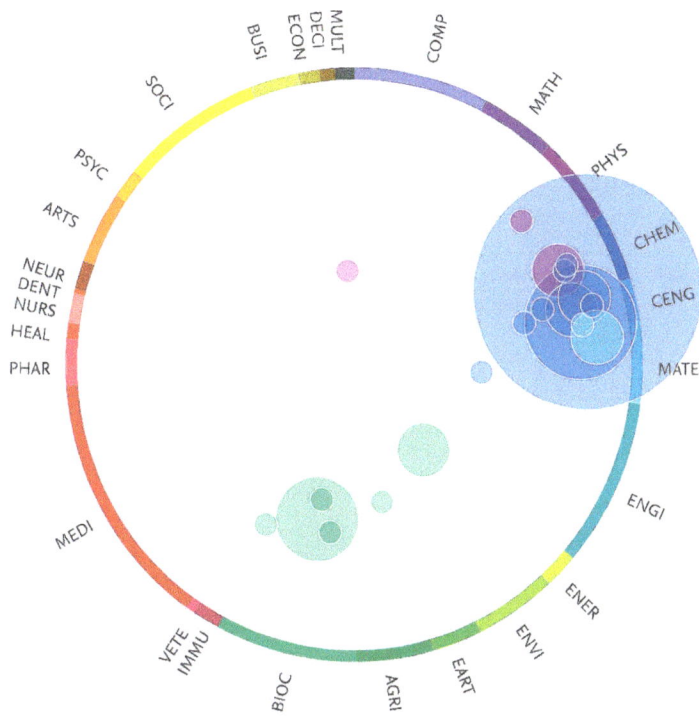

Figure 3. Visual representation of topics treated within this SI, as calculated by Scival prominence metrics in Scopus.

These skills are well interconnected. In addition, other areas of expertise at the boundary between biochemistry (BIOC) and biological sciences (AGRI), from the macro to micro scale, are growing and becoming linked together. Other skills, though, are new in the application to the conservation science field. This is the case of applied engineering (ENGI), which nowadays is a highly requested skill for numerical modelling for building or material component simulation (MATE), for the purpose of analyzing and testing new solutions in health monitoring or conservative interventions, as in the case of energy retrofitting solutions (ENER) in historical buildings. Other skills, with the potential to be applied in conservation science, are at an early stage of development or are less requested, as the case of pure climate simulations (EART) to estimate the impact of climate change outdoors/indoors, or mathematical models to assess the economic and social impact expected by climate change. This last topic, representative of adaptation to climate change and vulnerability assessment, is one that—at present—has both the highest prominence percentile (i.e., 99.5) and an increasing rate of interest. This means that the topics at the boundary between computer science/mathematics (COMP-MATH), economic/ econometrics (ECON), and decision science (DECI) are new and very promising topics that are likely to be well-funded in future calls for research grants.

4. Key Messages for Future Research

The wide portfolio of methodologies, data, and case studies presented in the contributions published in this special issue prove that heritage, material, and environmental science are currently vibrant research and practice domains, with expertise spread across the globe and teams fully exploiting the capability of innovative monitoring as well as experimental and numerical simulation techniques to investigate decay mechanisms on heritage materials and components, mainly on sites in different geographic and environmental contexts. It is clear that the intuition of heritage scientists, conservators, or experts in the field affects the assessment of climate change impacts on heritage materials, with respect to the analysis which a standardized evaluator could conduct. Experience sometime compensates for the difficulty encountered in estimating the exact decay rate, or in enlarging decay analysis to an extended family of materials and objects geometries still not considered in literature. Difficulties arise due to issues in quantitatively estimating natural ageing on materials, conservation–restoration treatments, or synergistic effects. These still unknown decay mechanisms, exacerbated nowadays by climate change, can be better understood through the collection of enough statistical data to be compared with experimental research and numerical simulations conducted on similar heritage materials and decay process to produce mathematical equations (i.e., damage functions). Specifically, long-term monitoring on standard environmental parameters, such as T, RH, and precipitation, as well as on data related to pollution, salt presence, and previous conservative treatments are the key to progress within the research.

What has also been reported in this SI is the interrupted flow of information between the theoretical knowledge available at the international level and the passing of that knowledge down to the local management scale. The lack of knowledge of management methodologies incorporating climate change impacts, as well as the need to identify and disseminate practical solutions and tools for mitigation, are tangible. To help heritage institutions adapt to a changing climate, or be effective in mitigation actions, less energy-intensive preventive conservation policies, evolving standards and guidelines over time, and a green-thinking approach to conservation should be implemented—for example, with the help of regulations and financial incentives. Finally, as support from the wider community is important for raising awareness and successful adapting to climate change, policies and initiatives to increase community engagement have to be realized.

Funding: This research received no external funding.

Acknowledgments: The guest editor would like to acknowledge all the authors for contributing to the special issue, as well as the anonymous peer reviewers for assessing the submitted manuscripts and greatly helping the authors enhance the scientific quality of their papers. Sincere gratitude goes to the Editorial Board and Office of *Geosciences*, especially to the managing editor Richard Li, for the invaluable help and assistance provided at all stages of the design, management, and publication of this special issue.

Conflicts of Interest: The authors declare no conflict of interest.

References

1. ICOMOS. *International Charters for Conservation and Restoration*; Monuments & Sites, ICOMOS: München, Germany, 2004; Volume I, ISBN 3-87490-676-0.
2. Council of Europe. Recommendation of the Committee of Ministers to member States on the European Cultural Heritage Strategy for the 21st Century. In Proceedings of the 1278th meeting of the Ministers' Deputies, Strasbourg, France, 22 February 2017; Available online: https://rm.coe.int/16806f6a03 (accessed on 29 May 2019).
3. Sabbioni, C.; Brimblecombe, P.; Cassar, M. *The Atlas of Climate Change Impact on European Cultural Heritage. ScieNtific Analysis and Management Strategies*; Anthem Press: London, UK, 2010.
4. Leissner, J.; Kilian, R.; Kotova, L.; Jacob, D.; Mikolajewicz, U.; Broström, T.; Ashley-Smith, J.; Schellen, H.; Martens, M.; van Schijndel, J.; et al. Climate for Culture: Assessing the impact of climate change on the future indoor climate in historic buildings using simulations. *Herit. Sci.* **2015**, *3*, 38–52. [CrossRef]
5. Anaf, W.; Leyva Pernia, D.; Schalm, O. Standardized Indoor Air Quality Assessments as a Tool to Prepare Heritage Guardians for Changing Preservation Conditions due to Climate Change. *Geosciences* **2018**, *8*, 276. [CrossRef]
6. Carroll, P.; Aarrevaara, E. Review of Potential Risk Factors of Cultural Heritage Sites and Initial Modelling for Adaptation to Climate Change. *Geosciences* **2018**, *8*, 322. [CrossRef]
7. Haugen, A.; Bertolin, C.; Leijonhufvud, G.; Olstad, T.; Broström, T.A. Methodology for Long-Term Monitoring of Climate Change Impacts on Historic Buildings. *Geosciences* **2018**, *8*, 370. [CrossRef]
8. Loli, A.; Bertolin, C. Indoor Multi-Risk Scenarios of Climate Change Effects on Building Materials in Scandinavian Countries. *Geosciences* **2018**, *8*, 347. [CrossRef]

9. Bylund Melin, C.; Hagentoft, C.; Holl, K.; Nik, V.; Kilian, R. Simulations of Moisture Gradients in Wood Subjected to Changes in Relative Humidity and Temperature Due to Climate Change. *Geosciences* **2018**, *8*, 378. [CrossRef]

10. Ciantelli, C.; Palazzi, E.; Von Hardenberg, J.; Vaccaro, C.; Tittarelli, F.; Bonazza, A. How Can Climate Change Affect the UNESCO Cultural Heritage Sites in Panama? *Geosciences* **2018**, *8*, 296. [CrossRef]

11. Fermo, P.; Goidanich, S.; Comite, V.; Toniolo, L.; Gulotta, D. Study and Characterization of Environmental Deposition on Marble and Surrogate Substrates at a Monumental Heritage Site. *Geosciences* **2018**, *8*, 349. [CrossRef]

12. Menéndez, B. Estimators of the Impact of Climate Change in Salt Weathering of Cultural Heritage. *Geosciences* **2018**, *8*, 401. [CrossRef]

13. Dotsika, E.; Kyropoulou, D.; Christaras, V.; Diamantopoulos, G. $\delta 13C$ and $\delta 18O$ Stable Isotope Analysis Applied to Detect Technological Variations and Weathering Processes of Ancient Lime and Hydraulic Mortars. *Geosciences* **2018**, *8*, 339. [CrossRef]

14. Sesana, E.; Gagnon, A.; Bertolin, C.; Hughes, J. Adapting Cultural Heritage to Climate Change Risks: Perspectives of Cultural Heritage Experts in Europe. *Geosciences* **2018**, *8*, 305. [CrossRef]

geosciences

MDPI

Article

Standardized Indoor Air Quality Assessments as a Tool to Prepare Heritage Guardians for Changing Preservation Conditions due to Climate Change

Willemien Anaf [1,2,*], Diana Leyva Pernia [3,4] and Olivier Schalm [1,5]

[1]   Conservation Studies, University of Antwerp, B-2000 Antwerp, Belgium; olivier.schalm@uantwerpen.be
[2]   War Heritage Institute, B-1000 Brussels, Belgium
[3]   Computer Science, University of Antwerp, B-2020 Antwerp, Belgium; diana.leyvapernia@uantwerpen.be
[4]   Department of Physics, CEADEN, 502, 11300 Havana, Cuba
[5]   Antwerp Maritime Academy, B-2030 Antwerp, Belgium
[*]   Correspondence: willemien.anaf@uantwerpen.be

Received: 6 July 2018; Accepted: 24 July 2018; Published: 27 July 2018

Abstract: Climate change will affect the preservation conditions of our cultural heritage. Therefore, well-considered mitigation actions should be implemented to safeguard our heritage for future generations. Environmental monitoring is essential to follow up the change in preservation conditions and to evaluate the effectiveness of performed mitigation actions. To support heritage guardians in the processing and evaluation of monitored data, an indoor air quality (IAQ) index for heritage applications is introduced. The index is calculated for each measured point in time and is visualized in a user-friendly and intuitive way. The current paper describes the backbone of the IAQ-calculating algorithm. The algorithm is subsequently applied on a case study in which a mitigation action is implemented in a church.

Keywords: indoor air quality; cultural heritage; climate change; preventive conservation; mitigation actions

---

1. Introduction

Climate change will not only change our way of life, it will also influence the preservation conditions of our cultural heritage. Increasing temperatures will cause a rise in sea level and changing precipitation patterns. Other expected effects of global warming are larger temperature fluctuations, an elevated risk of mold growth, more frequent salt deliquescence cycles, an accumulation of extreme weather events, such as heavy rain, flooding, droughts and strong wind, elevated UV-levels, etc. Moreover, the impact of growing (mass) tourism should be considered as well. On the other hand, a decrease in economic resources for heritage conservation is expected [1–4]. The expected impact of climate change on built heritage is already well-documented [4], but one should also consider the changing indoor conditions. This is certainly the case for historic buildings where the outdoor climate often highly influences the indoor conditions. To protect our heritage from these new conditions, adequate mitigation actions will be needed in the near future. Climate change occurs at a slow pace over several generations. Therefore, the adaptation of preservation conditions will occur as a consecutive series of (low-cost) mitigation actions that are sufficiently good for the time being, interspersed with some high-cost but drastic mitigation actions. For heritage guardians, it might be difficult to decide when and which mitigation actions should be implemented. An evaluation tool to assess the evolution of indoor air quality in a quantitative way could support the optimization of the series of consecutive mitigation actions and prepare heritage guardians for the change that is to come.

We introduce a standardized method that converts data streams collected with data loggers (i.e., the input) into a time series of indoor air quality (IAQ) indexes (i.e., the output). The index describes the overall air quality in relation to the preservation conditions of a specific material or object type. The use of air quality indexes, both for indoor and outdoor situations, is already widely used in environmental studies from other fields, especially those related to health impact and human comfort [5–9]. The IAQ index that we propose for cultural heritage applications is material specific and focuses on the indoor environment. It can be calculated for each measured data point in time, in contrast to time-averaged evaluations. Plotting the IAQ index over time by means of a line chart is a simple way to identify changes and trends in indoor air quality. It allows visualization of periods of elevated risk and the level of that risk and helps heritage guardians to identify hazards in a more focused way. Therefore, it offers a practical tool that supports decision-making towards the adaptation of the indoor environment to maintain certain preservation conditions despite climate change. Moreover, it can be used to objectively evaluate the effectiveness of a performed mitigation action.

The current paper describes the development of the algorithm that calculates the IAQ index from environmental measurements. The benefits of this approach are illustrated with a case study in which the effects of a mitigation action are shown.

2. Background

A well accepted method to follow up the preservation conditions of a heritage collection is to monitor an objects preservation state by regular visual inspections. The disadvantage of this method is that the hazards cannot be identified until there is visible damage. An alternative approach that enables early warning is based on the calculation of degradation rates of heritage objects from environmental measurements. The prediction of expected damage requires thorough knowledge of the relationships between environmental parameters and degradation rates. However, for many materials, the exact degradation mechanisms that describe that relationship are not yet fully understood. Alternatively, degradation rates can be predicted by (accelerated) degradation experiments under well-controlled conditions. The relationships between environmental parameters and the degradation rate are then described by a best-fitting mathematical function. Dose–response functions illustrate this approach. They enable the prioritization of the agents of deterioration and the definition of damage thresholds [2,10,11]. An example of an algorithm based on such mathematical functions is the preservation metrics developed by the Image Permanence Institute [12]. Unfortunately, dose–response functions are not available for all materials. Secondly, the degradation of a material is often influenced by the way it is integrated in the heritage object. Finally, the experimental conditions under which the functions are determined are not necessarily representative of natural conditions. Thus, the above approaches appear to be impractical for a generalized evaluation of the preservation conditions. Over the last two decades, risk assessments for collections have made their appearance in the heritage sector [13–15]. Such assessments estimate the risk towards a collection by considering the ten agents of deterioration. They tackle the following questions [16]: What might happen? How likely is that? What will the consequences be? Such risk assessments are often time-consuming and require relevant expertise. In this contribution, we propose an alternative risk-based approach that focuses on the indoor air quality for heritage preservation. This approach requires less expertise and is based on several easily applied principles that are validated through practical experience and theory. The following paragraphs describe the approach.

2.1. The Concept of Key Risk Indicators

From the huge amount of literature concerning the degradation of historic materials, it is possible to identify a large number of parameters that affect degradation rates. However, that reality is too complex to estimate the risk that damage might occur. Instead, we simplified it by using a first simple principle: the degradation rate of any material is, to a large extent, driven by a limited number of environmental parameters. This set of parameters can be grouped in four categories that correspond to the following agents of deterioration: incorrect temperature, incorrect relative humidity, radiation and pollution (Figure 1).

The small set of environmental parameters that dominate the degradation rate of all (historic) materials can be considered to be markers, i.e., distinguishing and easily measurable features that give an objective indication of the preservation state in which a collection resides. Well-known examples are temperature, relative humidity, illuminance and UV-radiation. If the risks caused by these markers are known, the overall picture of the preservation conditions is known. For that reason, the markers can be used to introduce the concept of key risk indicators (KRIs) [17–19]. KRIs are independent parameters that estimate the threat that certain preservation conditions will harm the collection. They rely on the measurement of a marker and on a corresponding description of the alarming situation where enhanced risk for accelerated degradation might occur. The following list gives an overview of the 12 most critical KRIs (i.e., type of threats) identified from the literature: too high relative humidity (RH), too low RH, too large RH fluctuations, too high temperature (T), too low T, too large T fluctuations, too high illumination, too high UV-radiation, too high concentration of oxidizing gases (O_3, NO_x, SO_2), too high concentrations of organic gases (acetic acid, formic acid, formaldehyde), too high concentrations of reduced sulfur compounds (H_2S, carbonyl sulfide (OCS)) and too high concentrations of dust ($PM_{2.5}$, PM_{10}, deposited dust) (Figure 1).

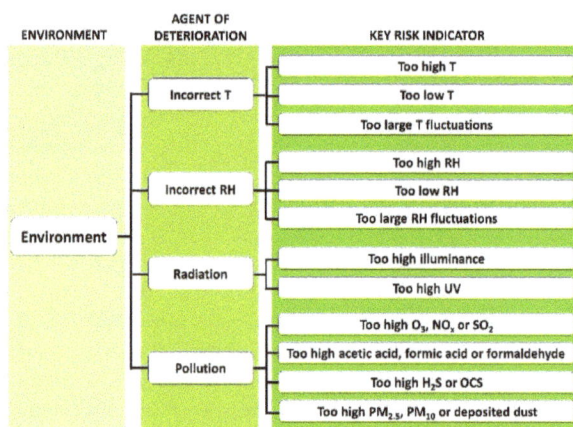

Figure 1. Schematic overview of the different levels by which the environmental appropriateness for heritage conservation are evaluated on. Abbreviations: RH, relative humidity; T, temperature; OCS, carbonyl sulfide.

2.2. Quantifying the KRIs

To simplify the estimation of the KRIs for specific environmental conditions, the question, "How fast do materials degrade?", is replaced by the question, "How large is the risk for enhanced degradation?". Although the answers of both questions contain similarities, they are not identical. For example, it is a complex matter to calculate the rate at which climate-induced damage accumulates in wooden objects from measurements of relative humidity and temperature [20–22]. However, we know that these parameters cannot be too low, too high or with excessive fluctuations without enhancing the risk of damage. This means that the level of risk as described by a KRI can be estimated by comparing the measurement of a marker with its corresponding target value. Such target values or ranges of acceptable values can be found in the literature, guidelines and standards.

The KRIs are quantified by converting their corresponding markers into a level of risk that is described by a value between 0 and 1—the higher that value, the higher the risk. Based on previous literature, four types of conversion functions have been identified. They are described in the list below and visualized in Figure 2. Since the shapes of the conversion functions are predefined, the exact definitions of the conversion functions are dependent on just a few nodes (i.e., the red dots in Figure 2, upper part). The position of the nodes coincides with published target values and is material-dependent. There is sufficient literature on thresholds, but their exact values are sometimes under discussion. In this contribution, one expert set these values and tested the results for consistency. The concept of calculating the level of risk with simplified conversion functions can be considered to be the second principle of the approach.

Conversion Function 1: This function describes the impact of the KRIs having a too high/too low RH or a too high/too low T. For example, for most hygroscopic materials, a mid-range RH has a limited risk of damage, while RH-values outside this recommended range are associated with higher risks. Materials for which a too low RH does not matter, such as metals, the first node is set at position (0,0).

Conversion Function 2: The fluctuation of a marker (e.g., RH or T) is defined as the maximum value minus the minimum value within a period of 24 h. Objects can usually withstand small fluctuations without damage. Therefore, until a certain magnitude of fluctuation, the level of risk for enhanced degradation is zero. The larger the peak-to-peak value becomes, the higher the risk is. From a certain peak-to-peak value, the risk for enhanced damage is so high that the level of risk is considered to be 1.

Conversion Function 3: This function describes the risk for enhanced degradation that is caused by the intensity of visible light and UVA radiation. At lower radiation levels, there is only a small risk of enhanced degradation, but that risk increases at higher intensities. At a certain intensity, degradation is almost certain to occur, and the risk becomes 1.

Conversion Function 4: This function describes the risk of all pollutant-related KRIs, i.e., oxidizing gases, organic gases, reduced sulfur compounds and dust. Although the exact influence of the pollutant concentration on the degradation of many materials is not known in detail, it is known that the lower the concentration is, the smaller the impact is (i.e., the ALARA principle: as low as reasonably achievable). A total of four nodes is used to define the conversion function, since well-accepted standards often mention a lower and a higher 'range' of threshold levels (e.g., reference [23]).

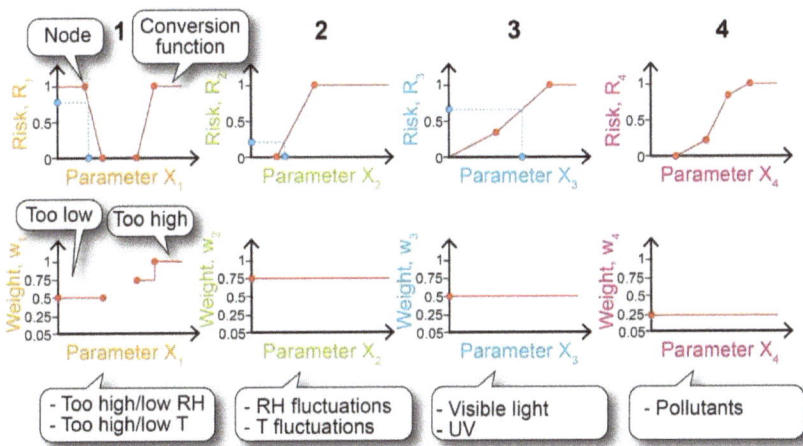

Figure 2. Conversion functions to calculate the level of risk that a marker is generating for a specific material or object type (upper part) and the way a weight is attributed to a key risk indicator (KRI) (lower part).

2.3. Risk Profile of a Material

The first principle states that the degradation rate of historic materials is driven by a limited number of markers. However, one single marker does not have the same effect on the degradation rates of different materials. For example, the same amount of radiation endangers very sensitive materials, such as paper and textiles and affects oil paintings to some extent, while metals are almost insensitive to it. On the other hand, when considering all KRIs on a single material, pollutants have, for example, a larger impact on metals than temperature. Therefore, the third principle states that weighting factors can be used (1) to rank the importance of the different KRIs per material or object type, and (2) to rank the sensitivity of material/object types per KRI.

A matrix was set up to elaborate the third principle. The matrix rows list 35 commonly occurring heritage materials and object types. Table 1 gives an overview of these materials and object types. They are considered to be representative for most heritage collections and cover materials and object types for which sufficient information on degradation can be found in literature. The matrix columns list the KRIs. First, the importance levels of the KRIs are ranked per material/object type (horizontal matrix direction). Five categories are allowed, and the same category could be attributed to several KRIs. The rankings are based on an extensive literature study, information from previous projects [24] and personal experience. Subsequently, the material/object sensitivity for each KRI was implemented using a five-category ranking as well (vertical matrix direction). To do so, the KRI importance within one material/object can change its ranking category, but the order of KRI importance within a material/object cannot change. Finally, the ranking categories are quantified by attributing a numerical score that reflects the impact of the KRI on the degradation: 0.05 (negligible), 0.25 (low), 0.5 (moderate), 0.75 (high) and 1 (extremely high). By using only five categories, disagreements between experts have a small effect on the final ranking because most disagreements are subtler that the rather broad categories that are imposed by our approach. The numerical scores are considered to be weighting factors.

Table 1. Overview of the commonly occurring materials and object types that represent most cultural heritage collections. They are classified in 14 main classes with the assignment of subclasses if relevant.

Material/Object Type	Subclasses
General collection *	
Paintings	Wood I Canvas I Copper
Paper	Cotton and rag paper I Groundwood containing paper I Lignin-free paper
Wood	Restrained I Unrestrained
Textile	Vegetable fibers I Wool/hair I Unrestrained silk I Restrained silk I Weighted silk I Synthetic fibers
Metal	Silver I Copper I Lead I Iron
Leather and parchment	Restrained I Unrestrained
Glass	General I Crizzling
Ceramic	Terracotta/earthenware I Stoneware/porcelain
Stone	Limestone I Gypsum I Alabaster I Marble
Ivory/bone/antler/horn	
Feather/insects/stuffed animals	
Photographs	Albumen I Collodion I Gelatin
Plastics	

* The material/object type 'general collection' offers an option that is material unspecific as a generic approach. If a sensitive object is present in the collection, one should opt to continue with this specific material.

In principle, the weighting factors describe the importance of each KRI. For this reason, the weight is independent of the marker value. Therefore, one weighting factor is assigned to each type of conversion function (Figure 2, lower part). The only exception is Conversion Function 1, because it combines two KRIs and they need to be weighted independently. Moreover, for the KRI 'too high RH', an additional weighting factor is attributed when crossing an RH of 75%. Above this value, an elevated risk towards mold growth can be expected. This additional weighting factor is only valid for mold-sensitive materials. In the range where the risk is zero, the weight is not defined because $w_i \times R_i$ remains zero.

For each material/object type, a spider graph can be plotted to visualize the relative KRI-importance. Each graph can be considered to be a risk profile for a given material/object type—the total area of the spider graph indicates the average sensitivity of the material/object to the overall preservation conditions. The differences in total area demonstrate that not all materials degrade at the same rate. Figure 3 gives an example for paintings, making the distinction between paintings on wood, canvas and copper.

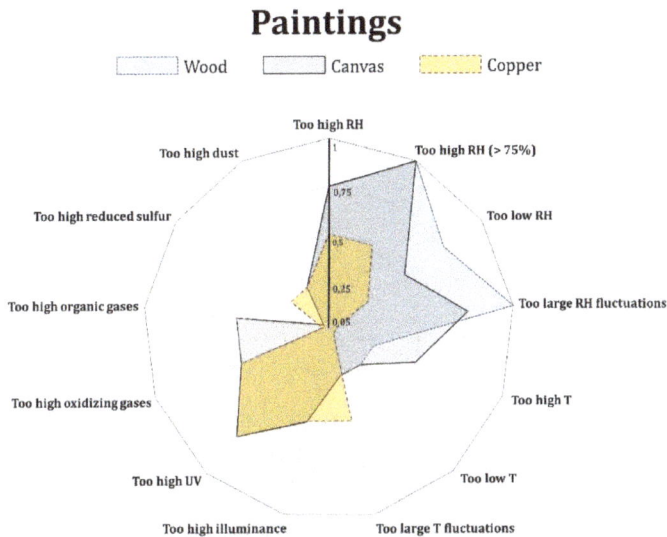

Figure 3. Spider plot with 13 dimensions to visualize the KRI importance for paintings on wood, canvas and copper. Five categories describe the impact on the degradation: negligible (0.05), low (0.25), moderate (0.5), high (0.75) and extremely high (1).

2.4. Combining all KRI into an Overall Indoor Air Quality (IAQ) Index

The preservation conditions are not determined by a series of marker-specific risks but by one overall risk. The IAQ index is related to that global risk. To calculate the index, the heritage guardian must first select which material/object type he wants to determine the indoor air quality for from a list of options. Then, the IAQ index is calculated with an algorithm that follows six subsequent steps (Figure 4), as follows: (1) The heritage guardian preprocesses the monitored environmental data to create a consistent data matrix to be uploaded. The matrix should be based on data of simultaneous measurements of markers at fixed time intervals. (2) Based on the material/object selection, the algorithm identifies which conversion functions are needed to calculate the level of risk for each KRI, R_i. (3) The algorithm now identifies the relative importance of the KRI based on the weighting factors, w_i. The levels of risk for the KRIs, R_i, are subsequently multiplied by the respective weighting factor, w_i. (4) The overall risk for a specific data point, R_{max}, is controlled by the highest weight-corrected marker-specific risk (i.e., max $\{w_1 \times R_1, w_2 \times R_2, \dots \}$). (5) Since risk is associated with the probability of occurrence of damage due to the preservation conditions, the probability that no damage will occur (i.e., the safety of the environment) is given by $1 - R_{max}$. This magnitude is defined as the overall IAQ index. The numerical value of this index varies from 0 to 1—the higher the index, the better the preservation conditions. The maximum value of the IAQ index is determined by the w_i of the marker that sets R_{max}. The algorithm is repeated for each data point, resulting in a time series of IAQ indexes. If needed, a marker-specific IAQ index can be evaluated as well, defined as $1 - R_i$. This marker-specific index does not consider the weighting factors. (6) The behavior of the IAQ-index over time can be visualized in line charts. Another visualization can be done by assigning a specific color to each IAQ value using a color map. This results in color bars that depict the IAQ index over time, allowing intuitive and user-friendly interpretation.

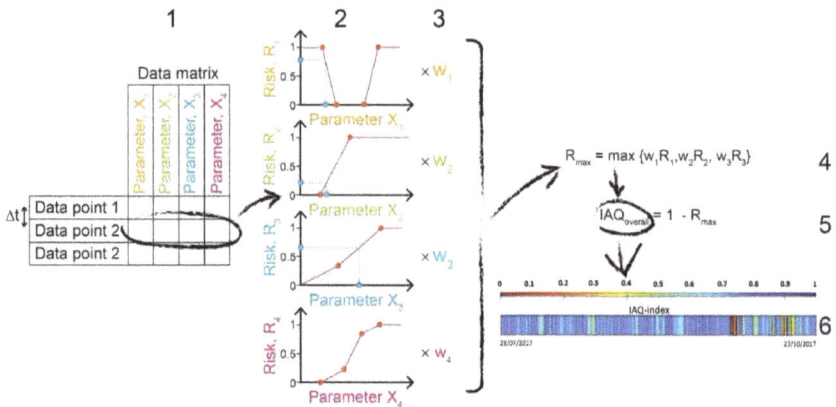

Figure 4. Schematic visualisation of the steps considered by the indoor air quality (IAQ) index algorithm.

3. Materials and Methods

3.1. Data Acquisition

An in-house developed multi-sensor tool measured a large number of markers. The monitoring tool consisted of a multi-purpose data logger (DataTaker DT85, Thermo Fischer Scientific, Scoresby Vic, Australia) to which a wide range of off-the-shelf sensors were coupled. The temperature and relative humidity were collected with a GMW90 (Vaisala, Helsinki, Finland). The intensity levels of visible and UV light were monitored with the upward positioned sensors SKL310 and SKU421, respectively (Skye Instruments, Llandrindod Wells, UK). Particulate matter was collected with a DC1100 Pro Air Quality Monitor (Dylos Corporation, Riverside, CA, USA). The measured concentration in number of particles m^{-3} was converted into $\mu g \ m^{-3}$ using an empirical formula provided by the supplier. Concentrations of NO_2 and O_3 were collected with NO2-A43F and OX-A431 sensors (Alphasense, Essex, UK). The concentration of total volatile organic compounds (TVOC) was measured with a photo ionization detector with a 10.6 eV lamp (Vaisala, Helsinki, Finland), but the concentrations were too close to the detection limit to get a meaningful signal. For other markers, such as H_2S, no appropriate mid-price sensor could be found that was able to measure the (low) concentrations.

All sensors were read out in phase with a frequency of 15 min and saved by the data logger. Data was downloaded wirelessly using a 4G router.

3.2. Data Processing

The collected environmental data were stored in a data matrix. The rows consisted of the timestamp and the series of sensor readouts. The rows were denoted as the measuring points. The measured markers were organized as columns. The IAQ index was calculated by following the procedure described above and using an in-house developed software written in MatLab R2017a (The MathWorks, Natick, MA, USA).

3.3. Data Visualization

Data visualization makes large data sets understandable and helps to absorb the information in a constructive way. Therefore, even though the IAQ index already summarizes the appropriateness of an environment based on several markers, an intuitive data visualization remains essential. The data should be visualized in a way such that it becomes useful and easy to understand by the end user, i.e., heritage guardians. It was decided to visualize the IAQ index over time using color bars. For this, the IAQ values were associated with the reverse jet color map from the software package MatLab R2017a (The MathWorks, Natick, MA, USA). By associating the IAQ index of each data point to a vertical colored line, the time series was converted to a color bar. A dark blue color was assigned to an appropriate environment, while a red color indicated an environment with elevated risk.

3.4. Sampling Location

The application of the IAQ index calculations was illustrated on a dataset collected in a late Gothic church in the centre of a small Belgian city. The church houses several canvas paintings, including a masterpiece of Rubens. Its other remarkable interior elements are a valuable organ, an early 17th century sacrament tower, several wooden statues, a wooden pulpit and metal candle holders. Environmental monitoring was performed at the organ loft at a height of around 7 m. Data collection started on 3 July 2017 and will continue until spring 2019. The current article focuses on a four-month period from 1 January 2018 to 30 April 2018. Within this period, a new heating system was started up in the church. The target temperature of the heating was set at 11 °C. When outdoor temperatures increase during the warmer seasons, the heating system is automatically switched off. For the IAQ index calculations, we selected canvas paintings, restrained wood and copper as the material/object types of interest.

4. Results

Figure 5 shows a traditional line chart for all measured markers in the church for the period, January to April 2018. At the beginning of January, temperatures in the church were below 10 °C, and correspondingly, high indoor humidity levels of around 80% were observed. On 25 January and 26 January, the newly installed heating system was tested. The first test resulted in a remarkable peak in particulate matter. This was linked to the resuspension of deposited dust in the heating grids. After the tests, the temperature increased from about 9 °C to 11 °C while the RH dropped from somewhat higher than 80% to 75–78%. The heating system was effectively put into operation on 1 February, with a target temperature of 11 °C. This resulted in a sudden drop in RH to 60%. Until the end of February, the temperature remained constant, but the RH continued to drop. It should be noted that there was a cold snap during this period, marked with outdoor temperatures below the freezing point and an outdoor decreasing RH as well. The combination of the outdoor cold snap and the indoor heating resulted in a continued decrease in RH down to almost 25%. Therefore, it was decided to lower the temperature set point of the heating system on 27 February. In mid-March, a failure in the installation occurred, resulting in a temperature drop. Subsequently, after a short period of heating, the outdoor temperature rose, and the heating system did not switch on anymore.

The other markers—illuminance, UV-radiation, NO_2, O_3 and $PM_{2.5}$—showed the presence of numerous peaks. The peaks in the pollution-related markers could not be related to the heating system, except for one PM peak on 25 January. Illuminance and UV radiation mainly showed day–night cycles. Approaching spring, the UV levels tended to increase relative to the winter period.

The graphs of all measured markers possess a wealth of information. However, when not familiar with data processing, the information can become overwhelming. The information output could be enhanced by adding yardsticks that denote the acceptable ranges as defined by guidelines. However, even with this information, heritage guardians could be lost in the data. To demonstrate the user-friendliness of the IAQ index calculations, the algorithm was run for this dataset for canvas painting, restrained wood and copper (Figure 6). When hygroscopic materials, such as canvas paintings and wood, are exposed for a long time in certain humidity conditions, it is expected that they will acclimatize to these conditions [25]. Therefore, for these materials, general RH threshold values for Belgian churches were considered [26]. Other threshold values were mainly based on ASHRAE [23], CIE [27] and Finney [28].

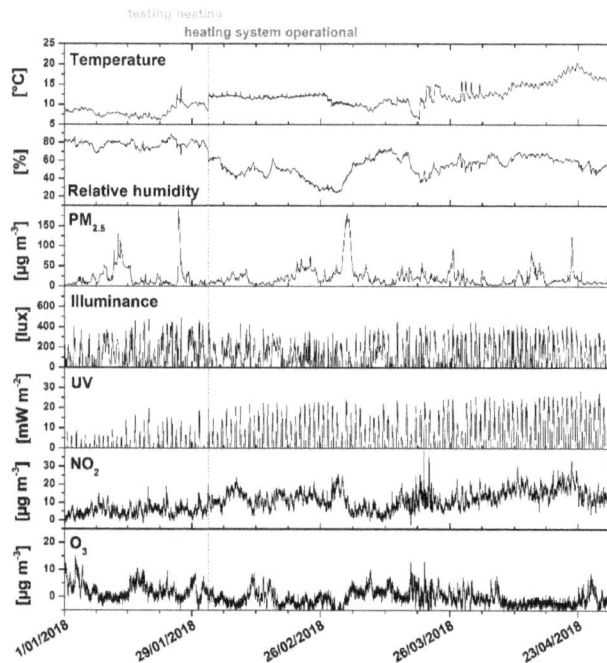

Figure 5. Scatter plot of the measured markers in the period, 1 January 2018 to 30 April 2018. The vertical dashed lines indicate the test period of the heating system and the moment at which the heating system became operational.

When comparing the overall IAQ index for these three material/object types, one quickly notices the correspondence between canvas painting and restrained wood, and the totally different outcome for copper. Indeed, canvas paintings and wood are both hygroscopic, while copper, as a metal, behaves in a different way. This results in a different sensitivity towards certain environments. For canvas paintings and restrained wood, a clear transition from a period with a high level of risk (dark red) towards more appropriate conditions (blue) can be observed around 1 February when the heating system was operational. After the heating system became operational, the IAQ index became worse again for a certain period. This period was more pronounced for restrained wood (orange color) compared to canvas painting (green to orange color). For copper, there is a rather equal IAQ evaluation throughout the whole period, with an intermediate IAQ index. To quantify the direct improvement of the start-up of the heating system, we considered the average IAQ index of one week before and one week after the heating system became operational. By considering such a short period in time, we focused on the short-term impact of this mitigation action, and eliminated other influences (e.g., seasonal change) and undesired situations as much as possible. The ΔIAQ between the weeks before and after the commissioning of the heating system equaled 0.6, 0.5 and 0.0 for canvas painting, restrained wood and copper, respectively.

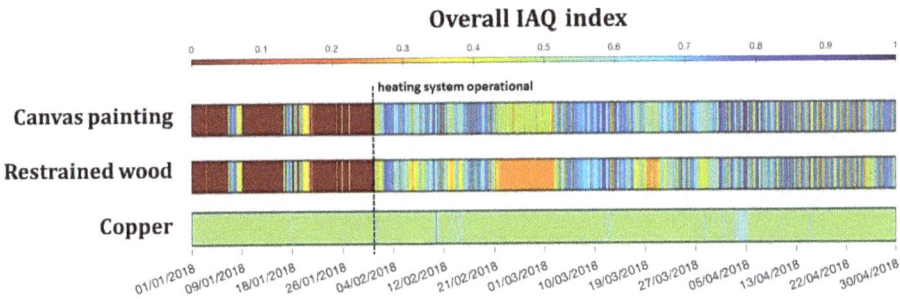

Figure 6. Overall IAQ indexes for canvas painting, restrained wood and copper over a period of 4 months. The dashed line indicates the moment at which the heating system came into operation.

The overall IAQ index gives a visual summary of the environmental appropriateness based on all (measured) KRIs. It compresses all information from the multiple environmental parameters of a data point into one single index. Based on the marker-specific IAQ indexes (1-R_i), the marker(s) that cause the undesired situation can be identified. Figure 7 shows an overview of all marker-specific IAQ indexes for restrained wood and copper, which are also visualized in color bars. For restrained wood, it is easily visible that the characteristic pattern in the overall IAQ index was mainly caused by a too high/too low RH. Temperature values and PM concentrations also exceeded the threshold levels, but these are estimated to have a lower impact on the general degradation rate of the collection.

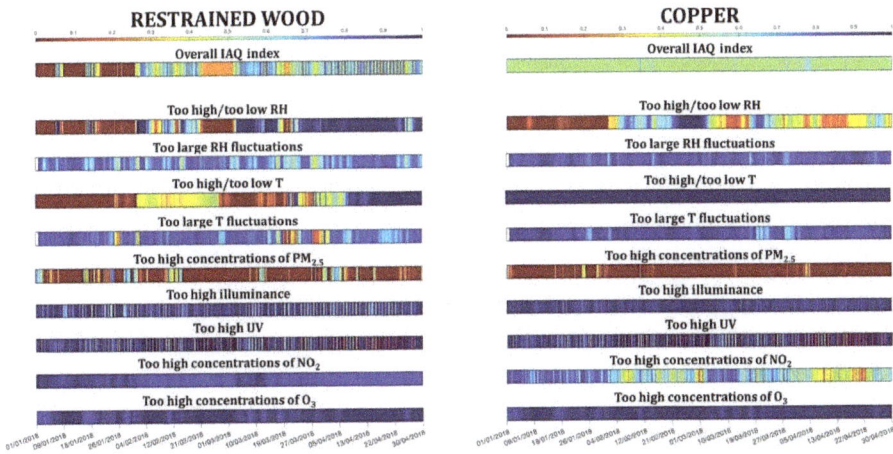

Figure 7. Overview of the parameter-specific risks for restrained wood (**left**) and copper (**right**).

For copper, the situation is more complex. The most striking undesired periods appeared for both a too high/too low RH and for too high concentrations of $PM_{2.5}$ (dark red). However, the overall IAQ index depicts colors in the greenish range, corresponding to IAQ indexes of around 0.5. The translation of the marker-specific IAQ indexes to the overall IAQ index seems counter-intuitive. This is due to following reason. Copper is not considered to be a highly sensitive material when compared to other material types such as, for example, paper. Therefore, the highest weighting factor attributed to any marker for copper is 0.5. This means that only half of the color scale for the overall IAQ-index is used. In this specific case, both RH and PM exceeded the thresholds. The weighting factor for both markers equaled 0.5. Since the IAQ index is calculated as the maximum of (w_1 x R_1, w_2 x R_2, ... , w_n x R_n), almost the whole period had an overall IAQ-index of 0.5. In addition, our approach evaluated the preservation conditions with several independent KRIs. Synergetic effects, if any, were not considered, for example, the increased degradation rate of dust particles in humid conditions. Therefore, the start-up of the heating system is not visible in the color bar of the overall IAQ-index but is reflected in the marker-specific indexes and, especially, in the color bar related to a too high/too low RH. This effect is a limitation of the algorithm: the choice of incorporating the sensitivity of materials relative to each other results in a loss of information. However, by evaluating the combination of the overall and the marker-specific IAQ indexes, the information can still be made available.

5. Discussion

Although several personal/human decisions (e.g., selection and interpretation of standards and guidelines, definition of weighting factors, etc.) are introduced in the algorithm, the algorithm itself is a standardized procedure that leads to reproducible and quantitative judgement of the IAQ. This standardized evaluator enables the conversion of measurements into judgements. These judgements help the heritage guardian analyse the preservation conditions and identify the hazards that are endangering his collection. Before using the algorithm, the heritage guardian is obliged to select the materials or object types for which he wants to know the IAQ index from a list of materials/objects. It is important to have measurements of all relevant KRIs for that material/object type. Current technology does not yet allow continuous measurement of all relevant markers with low-cost devices. However, with fast-evolving technology, it is expected that more sensors with better detection limits will become available. In the meantime, the algorithm can be applied, but one should be aware of the possible overestimation of the IAQ due to missing information of a relevant marker.

Since hazards can reoccur in the future with a possible increased level of risk, it is advised that mitigation actions are performed to avoid or reduce the identified hazards reoccurring in the future. This means that identified undesirable situations contain valuable information and should not be neglected, even when they have not caused any noticeable harm so far. With this approach, the slow change in hazard occurrence as a result of climate change will automatically be compensated for. The overall IAQ index can be used to detect periods of elevated risk. By looking at the marker specific IAQ indexes or the original line graphs, the causes of risk can be identified. By mitigating these risks, even small ones, the general preservation conditions improve, and material degradation slows. The algorithm can already be applied on short datasets and does not require a minimum of at least one year of data as an input. Therefore, it allows the environment to be followed up in real-time. When the preservation conditions start to become worse, one can quickly undertake action before (irreversible) damage occurs. In this way, the IAQ index serves as an early warning.

Despite the advantages and the user-friendliness of our approach, there are also some points of concern that should be considered. One limitation has already been described by the visualization of the overall IAQ index of copper where the ranking of the different KRIs based on their impact on one material and the ranking the KRI based on the impact of a series of materials resulted in a limitation in the visualization of all required information. Also, the following remarks should be considered when using the algorithm:

The intuition of the developer affects the definition of the standardized evaluator. The proposed principles do not perfectly reflect the complex reality and require some expert intuition. However, they work well enough to make several evaluations. The principles can be refined at a later stage, and the IAQ indexes of data from the past can be recalculated. Thus, the standardized evaluator generates reproducible and quantitative evaluations, but the scale is not absolute.

The exact degradation rate remains unknown. Although the IAQ indexes give good insight into the periods with elevated risk, the initial question, "How fast do materials degrade?", is not answered. However, the IAQ algorithm supports the formulation of that answer by estimating the enhanced risk for degradation. This already helps heritage guardians to make decisions.

Restricted options for material choice. The algorithm offers a list of 35 materials and object types to select. The list covers a wide range of heritage materials and objects that is representative of heritage collections. When using the IAQ index calculator, one should be aware that within each material/object type, variations in sensitivity exist depending on the applied techniques, material combinations, material purity, etc. These variations are one of the reasons why objects of art should be considered to be unique objects. Also the conservation state is important, since deterioration rates may vary during ageing [25], and conservation–restoration treatments can suddenly change the fragility of an object. Such refinements are not implemented in the algorithm which considers materials and objects at a statistical level (i.e., average materials and objects with an average behavior).

Synergistic effects. The IAQ index does not consider synergistic effects because it uses independent KRIs to estimate the risk of elevated degradation and not degradation mechanisms. Several synergistic effects are well known, e.g., lead corrosion is highly promoted in the presence of organic acids and high humidity [29,30]. Since synergistic effects are not considered, periods of elevated risk could be somewhat underestimated. This could happen when the ranking of the KRIs as described by the weights is affected. However, changes in ranking are only expected with strong synergistic effects. Even though the overall IAQ index could be underestimated in cases of such strong synergistic effects, the marker-specific indexes will still show the periods of elevated risk.

Evolving standards and guidelines. Due to improved knowledge and expertise, green-thinking, and less energy-intensive preventive measures, standards and guidelines for temperature and humidity tend to become more relaxed [31]. It is also expected that more accurate thresholds will become available for pollution levels. Therefore, to keep the algorithm up-to-date, the values that determine the conversion functions should be revised on a regular basis. Once revised, data from the past can be recalculated and re-evaluated considering the updated threshold values. This allows an evaluation of the progress in the IAQ despite changes in the 'standardized evaluator', i.e., the algorithm.

6. Conclusions

The IAQ-index estimates the indoor air quality in a standardized way. The proposed algorithm is a versatile tool that enables the introduction of a set of environmental parameters that goes beyond temperature and relative humidity. The introduction of an overall IAQ index and its visualization in color bars simplifies the analysis and interpretation of large data streams and a wide range of parameters that can be measured simultaneously. This offers heritage guardians a practical tool that helps them understand their indoor air quality. If more in-depth information is required, the user can consult the marker-specific IAQ indexes, or the original line graphs. The evaluation of such graphs becomes easier with the help of the IAQ indexes. This saves time and facilitates decision-making to improve the indoor air quality and to adapt the heritage institute to a changing climate.

All periods in time are evaluated in exactly the same way, resulting in a comparative risk of damage to heritage collections over time. Even if the following generations decide to adapt the algorithm, the data from the past can easily be re-evaluated and new comparative risks can be obtained. The comparative risk output also allows heritage guardians to quantify the effectiveness of mitigation actions and demonstrates that the effectiveness is not necessarily the same for all material/object types.

This substantiated and user-friendly tool will increase the awareness of a changing environment and will encourage heritage institutions to perform well-considered mitigation actions. In this way, the end-user will react to climate change-induced problems without necessarily realizing that these are related to climate change.

Author Contributions: Conceptualization, W.A., D.P. and O.S.; Software, D.P.; Visualization, D.P.; Writing—original draft, W.A.; Writing—review & editing, D.P. and O.S.

Funding: This research was funded by the Belgian Federal Public Planning Service Science Policy (BELSPO) under project number BR/132/A6/AIRCHECQ.

Acknowledgments: The authors thank the sponsors and the people that provided logistic support at the measurement location.

Conflicts of Interest: The authors declare no conflict of interest.

References

1. Saunders, D. Climate Change and Museum Collections. *Stud. Conserv.* **2008**, *53*, 287–297. [CrossRef]
2. Leissner, J.; Kilian, R. *Climate for Culture: Built Cultural Heritage in Times of Climate Change*; Fraunhofer MOEZ: Leipzig, Germany, 2014.
3. Huijbregts, Z.; Kramer, R.P.; Martens, M.H.J.; Schijndel, A.W.M.; Schellen, H.L. A proposed method to assess the damage risk of future climate change to museum objects in historic buildings. *Build. Environ.* **2012**, *55*, 43–56. [CrossRef]
4. Climate Change and World Heritage. Available online: http://unesdoc.unesco.org/images/0016/001600/160019m.pdf (accessed on 24 July 2018).
5. Lanzafame, R.; Monforte, P.; Patanè, G.; Strano, S. Trend analysis of Air Quality Index in Catania from 2010 to 2014. *Energy Procedia* **2015**, *82*, 708–715. [CrossRef]
6. Zhu, C.; Li, N. Study on indoor air quality evaluation index based on comfort evaluation experiment. *Procedia Eng.* **2017**, *205*, 2246–2253. [CrossRef]
7. Poupkou, A.; Nastos, P.; Melas, D.; Zerefos, C. Climatology of discomfort index and air quality index in a large urban mediterranean agglomeration. *Water Air Soil Pollut.* **2011**, *222*, 163–183. [CrossRef]
8. Murena, F. Measuring air quality over large urban areas: development and application of an air pollution index at the urban area of Naples. *Atmos. Environ.* **2004**, *38*, 6195–6202. [CrossRef]
9. Cairncross, E.K.; John, J.; Zunckel, M. A novel air pollution index based on the relative risk of daily mortality associated with short-term exposure to common air pollutants. *Atmos. Environ.* **2007**, *41*, 8442–8454. [CrossRef]
10. Strlic, M.; Grossi, C.M.; Dillon, C.; Bell, N.; Fouseki, K.; Brimblecombe, P.; Menart, E.; Ntanos, K.; Lindsay, W.; Thickett, D.; et al. Damage function for historic paper. Part I: Fitness for use. *Herit. Sci.* **2015**, *3*, 33. [CrossRef]
11. Strlic, M.; Thickett, D.; Taylor, J.; Cassar, M. Damage functions in heritage science. *Stud. Conserv.* **2013**, *58*, 80–87. [CrossRef]
12. Nishimura, D.W. *Understanding Preservations Metrics*; Image Permanence Institute—Rochester Institute of Technology: Rochester, NY, USA, 2011.
13. Ashley-Smith, J. Cultural Property Risk Analysis Model. Development and Application to Preventive Conservation at the Canadian Museum of Nature. *Stud. Conserv.* **2004**, *49*, 283–284.
14. The ABC Method: A Risk Management Approach to the Preservation of Cultural Heritage. Available online: https://www.canada.ca/en/conservation-institute/services/risk-management-heritage-collections/abc-method-risk-management-approach.html (accessed on 24 July 2018).
15. A Guide to Risk Management of Cultural Heritage. Available online: https://www.iccrom.org/wp-content/uploads/Guide-to-Risk-Managment_English.pdf (accessed on 24 July 2018).
16. Brokerhof, A.W.; Bülow, A.E. The QuiskScan—A quick risk scan to identify value and hazards in a collection. *J. Inst. Conserv.* **2016**, *39*, 18–28. [CrossRef]
17. Immaneni, A.; Mastro, C.; Haubenstock, M. A structured approach to building predictive key risk indicators. *RMA J.* **2004**, 42–47.
18. Taylor, C.; Davies, J. Getting traction with KRIs: Laying the groundwork. *RMA J.* **2003**, 58–62.
19. Scarlat, E.; Chirita, N.; Bradea, I.-A. Indicators and metrics used in the enterprise risk management (ERM). *Econ. Comput. Econ. Cybern. Stud. Res.* **2012**, *46*, 5–18.
20. Kozlowski, R. Numerical Modeling and Direct Tracing Experts Proceedings of the Roundtable on Sustainable Climate Management Strategies. In Proceedings of the Climate-Induced Damage of Wood, Tenerife, Spain, April 2007.
21. Jakiela, S.; Bratasz, L.; Kozlowski, R. Numerical modelling of moisture movement and related stress field in lime wood subjected to changing climate conditions. *Wood Sci. Technol.* **2007**, *42*, 21–37. [CrossRef]
22. Bratasz, L.; Harris, I.; Lasyk, Ł.; Łukomski, M.; Kozłowski, R. Future climate-induced pressures on painted wood. *J. Cult. Herit.* **2012**, *12*, 365–370. [CrossRef]
23. 2011 ASHRAE Handbook: Heating, Ventilating, and Air-Conditioning Applications. Available online: https://searchworks.stanford.edu/view/11842453 (accessed on 24 July 2018).

24. MEMORI. The MEMORI Technology. Innovation for Conservation. 2013. Available online: http://memori. nilu.no/ (accessed on 24 July 2018).

25. NBN. *Conservation of Cultural Property—Specifications for Temperature and Relative Humidity to Limit Climate-Induced Mechanical Damage in Organic Hygroscopic Materials*; Bureau voor Normalisatie (NBN): Brussel, Belgium, 2010.

26. Staatsblad, Omzendbrief ML/11 van 19 november 200 Betreffende de Kerkverwarmingen van Beschermde Monumenten. Available online: http://www.ejustice.just.fgov.be/cgi_loi/change_lg.pl?language=nl&la= N&table_name=wet&cn=2015120412 (accessed on 24 July 2018).

27. CIE, Control of Damage to Museum Objects by Optical Radiation. Available online: http://www.cie.co.at/ publications/control-damage-museum-objects-optical-radiation (accessed on 24 July 2018).

28. Finney, L. *Basic Conservation and Environmental Monitoring*; AIM Focus Papers: Edinburgh, UK, 2006; pp. 1–8.

29. Tétreault, J.; Cano, E.; Bommel, M.V.; Scott, D.; Dennis, M.; Barthés-Labrousse, M.-G.; Minel, L.; Robbiola, L. Corrosion of Copper and Lead by Formaldehyde, Formic and Acetic Acid Vapours. *Stud. Conserv.* **2003**, *48*, 237–250. [CrossRef]

30. Tétreault, J.; Sirois, J.; Stamatopoulou, E. Studies of Lead Corrosion in Acetic Acid Environments. *Stud. Conserv.* **1998**, *43*, 17–32.

31. Atkinson, J.K. Environmental conditions for the safeguarding of collections: A background to the current debate on the control of relative humidity and temperature. *Stud. Conserv.* **2014**, *59*, 205–212. [CrossRef]

geosciences

MDPI

Article

Review of Potential Risk Factors of Cultural Heritage Sites and Initial Modelling for Adaptation to Climate Change

Paul Carroll * and Eeva Aarrevaara

Faculty of Technology, Lahti University of Applied Sciences, Niemenkatu 73, 15100 Lahti, Finland; eeva.aarrevaara@lamk.fi
* Correspondence: paul.carroll@lamk.fi; Tel.: +358-4146-45092

Received: 3 July 2018; Accepted: 23 August 2018; Published: 29 August 2018

Abstract: There are a range of local weather- and climate-related factors that contribute to the degradation of cultural heritage buildings, structures, and sites over time. Some of these factors are influenced by changes in climate and some of these changes manifest themselves through a speeding up of the rate of degradation. It is the intention of this paper to review this situation with special reference to the Nordic Countries, where typical trends resulting from climate change are shorter winters and increased precipitation all year round. An attempt is made to initially draw up a classification of materials and structures relevant to cultural heritage that are affected, with a proposed numeric scale for the urgency to act. The intention is to provide information on where best to concentrate cultural heritage site preservation resources in the future.

Keywords: cultural heritage; preventative conservation; climate change; mitigation; adaptation; climate modelling

1. Introduction

There are a range of local weather- and climate-related factors that contribute to the degradation of cultural heritage buildings, structures, and sites over time. Some of these factors are influenced by changes in climate and some of these changes manifest themselves through a speeding up of the rate of degradation. The authors will be looking at existing literature from relevant geographic areas, especially concentrating on the models proposed, which contain a promising methodology aimed at contributing to the body of knowledge to apply to preventative conservation in Finland. The emphasis will be on built heritage and structures constructed from all materials most commonly used in the Nordic Countries.

Researchers in this multi-disciplinary area usually tend to concentrate on particular aspects of either the mechanisms of climate change, or its effects on particular material groups. While it is the intention here to refer to a range of these, the emphasis will be on approaches relevant to Southern Finland, which have either a sub-arctic or warm summer continental climate (according to the Köppen system). In this area, typical trends resulting from climate change are shorter winters, warmer summers, and increased precipitation all year round. While several of the studies referenced occurred under different climate zones, the adaptability of the methodology, and thereby the relevance of the results of these to the region of special reference in the present study, will be reflected on in each case. Our interest in cultural heritage sites is also connected with the UNESCO (The United Nations Educational, Scientific and Cultural Organization) global geopark concept, in which the valuable geology, cultural heritage, and liveability of the area are reviewed together. This perspective requires not only the preservation and sustainable use of the geologically valuable formations, but also the need to promote the maintenance and sustainability of the cultural built tradition in the area, together with the development of geotourism and environmental education [1]. Visitors to geoparks in this network expect to experience both natural geological features in their natural form, as well as cultural heritage existing as a result of human activities, reflecting the special features of the region and how people through time have used the local natural features of their region to make a living and express themselves.

UNESCO [2] distinguishes between two types of tangible cultural heritage: there are moveable (paintings, sculptures, coins, and manuscripts) and immovable (monuments and archaeological sites) types, and then also those of the underwater variety. Ethnography divides cultural tradition into material (buildings, artefacts, etc.) and immaterial (culture, customs, ceremonies, storytelling, music, etc.). This study will deal with the immovable type of cultural heritage above the surface.

Research dealing with climate change impacts on cultural heritage has gathered more attention over the past decades. The FP6 project Noah's Ark was carried out from 2004 to 2007 and the project produced a Vulnerability Atlas and Guidelines for cultural heritage protection against climate change. The scientific report introduced the vulnerability risks due to extreme weather events typical of climate change. Further research was prepared under the FP7 Climate for Culture Project during 2009–2014. The focus of this project was to evaluate the slow ongoing impacts of climate change, instead of extreme events. The report also included the prediction of sea-level rise impacts on coastal cultural heritage. The ongoing project HERACLES—H2020 Heritage Resilience against Climate Events on Site (2016–2019)—will promote solutions or systems for effective resilience with a multidisciplinary approach. Another H2020 project STORM (2016–2019)—Safeguarding Cultural Heritage through Technical and Organisational Management—is collecting a set of non-destructive methods of surveying and analysis to enhance better predictions for the future climate change impacts on cultural heritage. The project is carried out in collaboration with ICCROM (International Centre for the Study of the Preservation and Restoration of Cultural Property). Several other research projects are working with similar research goals and targeting certain parts of Europe and the most typical threats in their regions [3].

2. Potential Risk Factors to Cultural Heritage

The range of elements in local weather that are being and will be potentially altered as a result of global level climate change is not always agreed on, but there is general consensus on those which could be damaging to aspects of cultural heritage. Brimblecombe [4], referring to the increasingly damp English climate, lists the following five: 1. Rainfall; 2. flooding and soil moisture content; 3. extreme weather (winds and rainfall); 4. temperature and relative humidity; and 5. pests and diseases (humidity and temperature affect pests). Humidity prevents wooden and brick buildings from drying during certain periods of the year, leading to structural stress.

Lemieux et al. [5] refer to Last Chance Tourism (LCT), but interpret the increased interest in it as being based on perceptions of climate change and specific aspects, such as glaciers melting and icecap ice retreating; the central idea of the concept being that people are motivated to visit places while they still exist in the present form. Forino et al. [6] give a longer list of nine categories of climate change-related impacts on cultural heritage, referring specifically to Australia. This paper first generally summarises them as fitting into the three categories based on how their effects come about, which are: meteorology, hydrology, and climatology. Here, meteorology refers to different types of storms, and climatology to extremes of temperature. The longer list itemises: 1. various types of physical damage; 2. soil instability; 3. susceptibility to changing soil moisture; 4. changes in hydrology; 5. changes in humidity cycles; 6. changes in vegetation; 7. migration of damaging pests; 8. climatic zone movements impacting cultural landscapes; and 9. changing economic and social patterns of settlements.

Philips [7] interviewed a range of professionals involved with UK World Heritage sites in order to assess, firstly, how climate change was taken into account in preservation plans and, secondly, how those working with these sites are informed or react to the information provided to them on climate change considerations. Since 2006, it has been a requirement for the management of sites in this network in the UK to have not only an appropriate management plan for their protection for future generations, but specifically one for dealing with the possible impact of climate change. The range of reactions obtained from the interviewees fitted into five categories: not knowing what information they receive is relevant or reliable; uncertainty about the whole science of climate changes; a lack of availability of the skills or knowledge needed to act; difficulties with risk perception and getting others motivated about climate change issues; and lastly, the challenge of devoting additional time or financial resources to cope with climate change challenges. Accordingly, she observed that the obstacles identified in her study as a whole have slowed down climate adaptation in the management of UK World Heritage sites, and that the mainstreaming of such considerations is required and will be able to happen if climate change impact is not dealt with as a separate issue of its own, but rather within the context of other risk preparedness.

In another publication, Philips [8] introduces the concept of the adaptive capacity of cultural heritage under the impacts of climate change. Adaptive capacity is an approach to investigating the state of the management of cultural heritage sites in consideration of climatic change. The key determinants of adaptive capacity are defined: as 1. learning capacity, 2. room for autonomous change, 3. access to resources and, 4. leadership in an institution. The qualitative research material was gathered from persons engaged with the management of different heritage sites in the UK that had suffered from severe weather event impacts in the previous five years. Based on the results obtained, the concept of adaptive capacity was divided into six different factors: resources, access to information, authority, cognitive factors, learning capacity, and leadership. This concept creates a wider framework in which risk assessment is an important factor and must be combined with other competences to successfully manage cultural heritage under climate change.

According to Kaslegard [9], increasing strain will affect buildings with cultural value. A warmer and damper climate will cause the deterioration of building materials. Coastal buildings face the impacts of sea level rise, flooding, and erosion. In general, more extreme weather events will cause more acute damage to traditional buildings. Due to climate change impacts, the management of cultural heritage will face new challenges. It is suggested that more attention will need to be paid to the identification, documentation, and mapping of those heritage sites that are most vulnerable from the point of climate change. More intensive maintenance and coastal defence methods will be needed as future actions.

Arctic regions are predicted to face the greatest increase of warming in the winter time. The temperature is expected to rise by 3–4 °C by 2050. The rainfall in Nordic countries will increase by about 10% on an annual level. Exceptionally, the west coasts of Norway and Finland might face as much as a 20–30% increase in rainfall in winter periods. At least extreme rainfall events are expected to appear more frequently in the whole Nordic area in the future. For example, in Norway, two out of three cases of building damage are classified as having been caused by humidity affecting the outer surfaces of buildings, such as roofs, facades, and floors in contact with the ground. All kinds of building materials suffer from high humidity, while in Nordic countries, the building tradition is mostly based on timber constructions. Impacts of rainwater and meltwater are also identified in wooden buildings, causing favourable conditions, for example, for fungal growth, different pest damage, and biological growth, such as mosses and algae, while water penetrates into the building through different surfaces.

Physical disintegration means the decomposition of materials into smaller fragments; that is typical, for example, with traditional brick buildings. In practise, this can be caused by frost damage or salt crystallisation, which both cause damage to the appearance of the building. The character of frost damage can be described through the freeze/thaw cycle, meaning the phenomenon when the temperature falls below zero and then climbs to over 0 °C again. Considering both freeze-thaw cycles and wet frost, most parts of Finland, the inner and northern part of the Scandinavian Peninsula, and the Arctic regions are likely to face greater risks of frost damage, although the risk will remain at a moderate level. The salt crystallisation is likely to grow in all Nordic countries because the increase of humidity brings the salt out from the built structures more easily.

Cultural heritage buildings, especially old industrial buildings, but also other types of buildings, contain metal building elements like iron beams, iron bolts, and wall anchorages in stone or brick walls, as well as roofing and guttering of copper or zinc. Chemical decomposition causes corrosion in these kinds of structures and might also affect the stone and brick structures of the building. Corrosion, together with humidity and temperature, is a threatening combination. Although, it has been noted that the occurrence of acid rain due to SO_2 pollution is not currently as serious a problem as it used to be in earlier decades [9]. The blackening of building facades was earlier caused by industrial processes. However, due to environmental legislation, this kind of poor air quality is no longer a problem, but instead of that, the building facades are suffering from emissions consisting of organic-rich pollutants (like fine carbonaceous particles) from vehicle exhausts which can also cause a change in the colour of the façade, especially in buildings constructed of calcareous stone or other porous material [3].

La Russa et al. [10] have described how the environmental impact on cultural heritage can be observed through the formation of black crusts on stone, which appears to be principally due to airborne heavy metals formed through combustion. Although more relevant to urban environments and not resulting from climate change, the X-ray spectrometry analytical methods described in this study provide information that can help in better protecting the structures in question.

Cultural heritage will face several threat factors caused by climate change, but one can also conclude that the traditional building materials and structures have capabilities to recover more easily from heavy rain events and flooding. Usually, the structures are more permeable, which ensures natural ventilation and helps with drying out [9].

3. Modelling for Mitigation and Adaptation to Climate Change

Forino et al. [6] presented the *Cultural Heritage Risk Index* (CHRI), which gives a score from 1 = next to no risk to 10 = the greatest risk of loss of cultural heritage assets. It is designed to be applied to particular sites and first takes three categories of analysis: hazard analysis, exposure analysis, and vulnerability analysis. The findings from these are then combined and subjected to a risk analysis to give the result. Hazard analysis is stressed more in scoring, with the ratio between the three initial analysis types being 5:3:2.

As part of the EU research project Climate for Culture [11], an innovative method for assessing climate change impact was proposed and described, but applied it to a specific aspect of cultural heritage, namely to wooden buildings. Nevertheless, the methodology applied contains a number of aspects that are relevant and transferable to other materials. Here, the Regional climate Model REMO, which was developed at the Max Planck Institute for Meteorology [12] was used, to give a more accurate set of predications for future climate conditions.

Modelling of different kinds of environmental hazards is also connected with disaster risk reduction (DRR) and disaster risk management (DRM) in order to reduce the impacts of different disasters, like severe climate change impacts. In 2005, the Hyogo Framework for Action (HFA) 2005–2015 was introduced as the first international model for Building the Resilience of Nations and Communities to Disasters (Hyogo Framework) [13]. The strategy launches five different priorities of action, like national and local priorities connected with institutions, the use of knowledge, innovation, and education to support culture and resilience, reducing the underlying risk factors and strengthening of disaster preparedness at all levels to minimise the impacts. Priority action 2 concentrates on identifying, assessing, and monitoring disaster risks and enhancing the early warning of them. These viewpoints can be successfully adapted for the modelling of and adaptation to climate change.

Romãoa et al. [14] concluded that a framework adaptable to risk assessment in the built environment needs to involve the following viewpoints: (1) reliable and sufficient data to establish suitable hazard models; (2) sufficient and reliable data on the assets under risk; (3) suitable procedures to model the vulnerability; (4) adequate models to predict the multidimensional consequences of the hazardous event; and (5) sufficient human, time, and economic resources. It is recognised that there is often a lack of adequate models to predict impacts, as well as a lack of sufficient resources. They also conclude that it is important to have a simple methodology adaptable to preliminary risk analysis, especially in the case of cultural heritage. A qualitative approach is suggested as a usable method to evaluate multidimensional risks in the cultural environment, due to its complexity. Understanding the behaviour of cultural heritage sites and structures is more important than detailed measurements of the objectives when developing a simplified model for practical use.

One of the principal materials of interest in this study is stone. Hambrecht & Rockman [15] point out that although stone is considered a very strong material and one that is resilient in the face of climate change, that this is not always the case; they refer to several study projects, such as that of Goudie [16], that reveal the worrying vulnerability of stone under the influence of variables such as moisture and vegetation. The type of rock is another relevant factor and softer stones such as limestone and soapstone are eroded much more quickly [17]. While the predominant stone type in Fenno-Scandinavia is hard granite, there are also natural occurrences of softer stone types.

Gomez-Heras and McCabe [18] go so far as to propose that by studying past environmental changes on stone structures, it can be possible to predict potential future impacts. They add that although the weathering of such structures is similar to that of natural stone as such, we can still get more information about anthropogenic impacts from the stone that has been used by humans at some stage in history. Stone weathering as an indicator of climate change offers interesting potential for future study and its incorporation as a tool in further models for working with cultural heritage protection.

The model being proposed by the authors of the present paper (the main elements of which are combined in Table 1 below) is intended to serve the purpose of prioritising cultural heritage elements and sites for protection against the ravages of climate change and therefore to help in the planning of adaptation and/or mitigation steps.

Table 1. Causes, results, and proposed level of urgency for acting, with comments on relevance to the case study.

Climate Change Category	Measure or Scale	Result/Effect	Materials/ Structures Affected	Proposed Urgency Rating *	Case Study: Application of the Criteria
Warmer Climate	Rise in °C/year	Freeze–thaw damage	Stone Brick	3	Partly visible in stone constructions (cow house), although the structure is also affected by the current use and site conditions.
		Rust	Metal	5	Limited use in case study buildings. Non-painted roofs are suffering from rust.
		New fauna–pests	Wood Brick	5	Clearly visible increased effect, especially in wooden facades, but also brick facades.
Longer Growing Season	Days/year	New/increased flora, algae, moss, root damage	Wood Brick Stone	5	Clearly visible increased effect, especially in wooden facades, but also cement surfaces like staircases and foundations (moss).
Increased Precipitation: rain or snow	mm/year	Humidity	Wood Brick Structures	10	Clearly visible effect in all facades, especially in northern and shaded facades and wooden building parts.
		Increased loads (snow)	Wood Brick Roof/Roof Structures (Typically Wood)	5–10	Depending on the roof material and declination; the lower the declination, the higher the risk of damage due to the increased load.
		Soil and material degradation	Foundations Base Floor	5	The highest risk with the buildings situated on slopes of the site (combination of different soil types).
		Flooding (from any increased precipitation effect)	Wood Brick Structures	10	The highest risk with the buildings situated on slopes of the site (surface runoff gathering).

Table 1. *Cont.*

Climate Change Category	Measure or Scale	Result/Effect	Materials/ Structures Affected	Proposed Urgency Rating *	Case Study: Application of the Criteria
Severe Rain Incidents	mm/hour	Erosion	Wood Brick Stone	5	The highest risk is with the buildings situated on slopes of the site (possible soil erosion).
Extreme Winds	m/s	Damage to structures through falling trees or wind causing damage to the roof	Metal Roofs Wood & Brick Structures	5–10	High steel roofs facing the dominating wind direction are subject to the largest threat of damage from extreme winds (residential building, cow house's high roof).

* *A key to the numeric scale is given in Table 2 below.*

Table 2. Proposed scale for urgency to act with particular materials/cultural heritage sites.

1	A mild or minor perceivable long-term effect (100 years or more)
3	A major perceivable long-term effect (50–100 years)
5	A mild or minor perceivable short- to mid-term effect (1–50 years)
10	A major short- to mid-term effect

Then, by combining elements of the modelling methods outlined above, and illustrated in summary in Figure 1, the following scoring system from 1–10 (described in Table 2 above) is suggested as a means of evaluating the relative risks to different materials. The score in question can apply to either individual materials or to an entire structure or site, and takes into account both the vulnerability of the item in question and over how long a timescale potential damage to it may have an effect, as well as how severe this effect is likely to be.

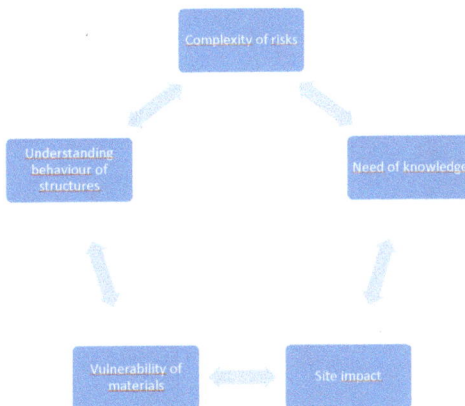

Figure 1. Elements having an impact on a practical-based model.

4. Case Study—Model Application

As a case study, we present a Finnish farmhouse complex consisting of several buildings, the map location and aerial views of which can be seen in Figure 2a,b below respectively. The built environment is classified as part of a regionally valuable village and cultural landscape [19]. The residential building is built in several parts: the oldest part is a log construction from the 19th century and a two-storey addition was constructed in the 1930's with wooden frames. The façade consists of horizontal wooden panelling with double glazed windows. The residential building is situated on a hill and other buildings are on lower positions of the site or on slopes. Most of the storage buildings are also horizontal log structures, except for a cow house built of cement brick in 1930. The roof materials are mostly steel and the largest storage hall has cement brick tiles. Similar traditional farmhouse complexes are quite usual in Finnish rural areas, and sometimes animal buildings are constructed of ordinary brick or granite stones. Natural granite stones have been used in the foundations of the residential building, but the stone surface has been covered with cement at a later phase. In the above table, the case study buildings and site are evaluated according to the proposed model and conclusions made dealing with the adaptability qualifications.

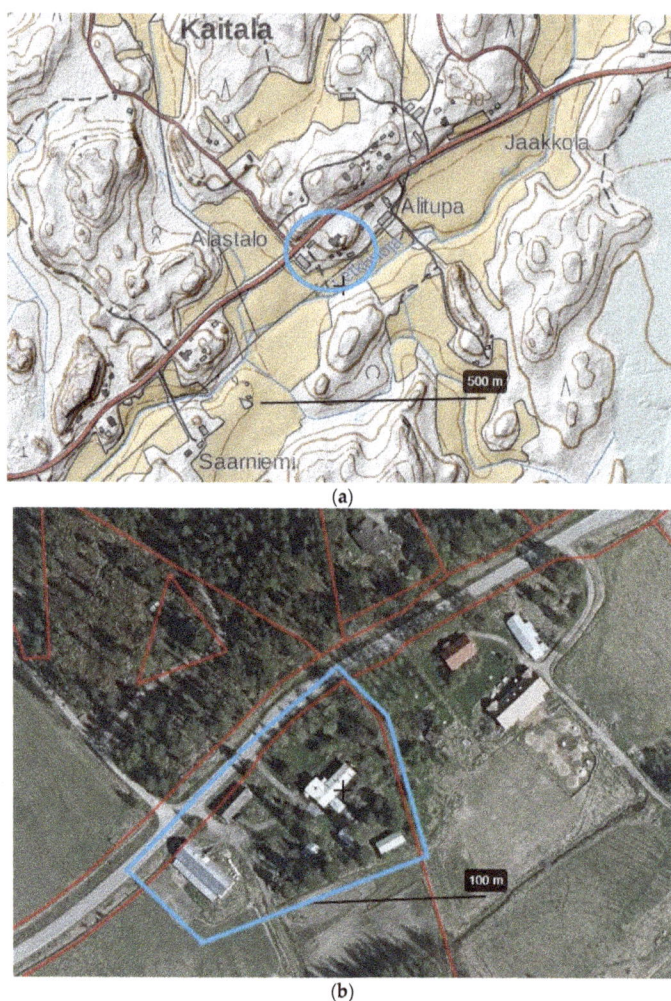

(a)

(b)

Figure 2. (a) The case study farm location in the village [20]. (b) Location of the case study buildings at the site [20].

5. Comments and Conclusions Based on the Case Study

The observations of the climate change impacts are based on the professional experience of an architect (one of the co-authors) living in the building complex, designing renovations and maintaining separate buildings, and following the changes caused by climatic conditions.

In this case, study wood is the dominating material of separate buildings. Natural stone has been used together with a cement surface, mostly in the foundations of the buildings. An exception is the big cow house built of cement tiles, which are not as durable as burnt tiles but have mostly survived in a moderate condition for almost ninety years. Cement tiles were fabricated in situ and the quality of the sand used in them had a significant impact of the total quality of the material, which is also visible in different parts of the building. In wooden buildings, the facades need repair and repainting regularly. The maintenance has not been very intensive, but the facades are mostly in a moderate condition. It has been clearly noticed that wooden facades are suffering more and more from the increased humidity, because they get dirty due to different kinds of flora and moss starting to grow on them. They need primarily cleaning with a limited use of water and suitable chemicals, instead of repainting.

In the case study, the conditions of the site also have a significant impact on the buildings. While the residential building is situated on a hill, this position protects it from severe impacts of flooding. The other buildings situated on slopes of the site are more sensitive to suffering in some ways from severe rain events and run off, as well as erosion. The site is sometimes subject to strong winds, especially those coming from the southwest, which is the dominating wind direction in Southern Finland. There has been damage caused by the strong wind tearing away parts of the roof in the highest buildings with steel roofs.

It has to be mentioned that it is essential to separate different types of damage in traditional buildings. Some of them can be caused by the wrong repair and renovation methods, like the use of unsuitable materials and paintings in old structures and surfaces. The use of different kinds of plastics and unsuitable insulation materials has caused damage to old structures when humidity is blocked inside them, causing, for example mould. In this case study site, there have not been unsuitable materials used, and the problems discussed are connected to climate change impacts.

6. Final Discussion and Conclusions

In predicting and planning to combat the detrimental effects of climate change on cultural heritage sites, we are faced with many unknowns: it is not clear by what scenario climate change will proceed, meaning somewhere from the optimistic slower rate to a pessimistic more rapid and intense one. Projections are currently being generated by different Global Climate Models (GCMs) hosted by different research institutions, each with its own formulation of the atmospheric flow dynamics and physics.

Four Representative Concentration Pathways (RCPs) have been considered in the fifth Assessment Report (AR5) of the Inter-Governmental Panel on Climate Change (IPCC) [21]. These Greenhouse Gas (GHG) concentration (not emission) trajectories, all considered as realistic, are used by modellers as atmospheric system forcing for generating climate response and change projections. The RCPs, namely RCP2.6, RCP4.5, RCP6.0, and RCP8.5, have been defined according to their contribution to atmospheric radiative forcing in the year 2100, relative to pre-industrial values. RCP4.5 is based on active GHG emission reduction interventions that could lead to a ceiling of approximately 560 ppm CO_2 (a doubling of atmospheric concentrations since the start of the industrial revolution) by the year 2100, while concentrations could stabilise or even decrease after the year 2100 [21] The RCP4.5 and RCP8.5 trajectories are associated with CO_2 concentrations of approximately 560 ppm and 950 ppm, respectively, by the year 2100 [22].

The local weather changes will also vary in terms of amounts of precipitation and extremes of temperature, storms, and so on; then, the predictions for how various materials (here, in particular, all traditional materials that are typical elements of Nordic cultural heritage: wood, brick, and stone) will react, contain uncertainties. Nevertheless, it is still possible to at least initially classify these risks using indexing methods, and then based on what the risks are perceived to potentially be, models can be drawn up for how to deal with them and as a whole serve to reduce the levels of uncertainty involved. It is important to collect more case examples to evaluate in practise the climate impacts on cultural heritage in different regions and places, test the models, and draw conclusions on the basis of the case studies. More advanced monitoring systems and regular evaluations will help to share the findings with researchers, professionals, and common users.

Combining ideas from the work described above on the potential for using stone as an indicator for the effects of climate change on stone [18] and the field experiments carried out by Daly [23] in Ireland, where small cubic samples of various types of stone were exposed to the influences of the weather for defined periods, there could be very useful results obtained in a future controlled experiment, for example, in the Nordic climate context such as that of Finland. This could potentially involve a set of samples of different kinds of stone materials exposed to the elements throughout all weather extremes of the year. Combined with observations from the past of the weathering of stone, the new data obtained from such a field test could reveal valuable information about the usefulness of stone as an indicator and ultimately about the effects of climate change on cultural heritage involving stone.

It can be concluded that the natural processes which lead to the deterioration of cultural heritage features will, as a whole, increase in their effect as a result of climate change; there will also be new potential threats involved. Many unknowns remain as to the relative influence of certain factors on particular materials, and although predicting these is difficult, there is a benefit of using models and attaching them to different scenarios for how severe climate change could be. Furthermore, an awareness of the relative urgency to react to threats to different building materials and types of structures can help in planning to protect them in time. Although the numeric scale for this purpose provided above, as well as the respective classifications into climate change threats and material features, all contain approximations, it still may, as a whole, contribute to the planning process towards the mitigation of the effects of climate change on cultural heritage, and to the appropriate adaptation to these changes.

Even if risk analysis provides important information about the possible occurrence of damage involving cultural heritage, the vulnerability should also be regarded as one aspect in a larger context of management of cultural sites and buildings. The preconditions of a successful management system are being able to secure the preservation of cultural heritage and to identify other actions needed to prepare and mitigate for the damage likely caused by changing climate conditions.

Author Contributions: This paper was entirely conceived, designed and written by the two co-authors E.A. and P.C. in a good spirit of cooperation. While E.A. suggested much of the cultural heritage background and provided the figures and the case study data and related interpretation, P.C. carried out the greater part of the review of previous research results and the basis of the central model together with the related tables, as well as the general conclusions.

Funding: This research received no external funding, but was enabled through the participation of the organisation in Finland, where the two authors are employed; Lahti University of Applied Sciences, Faculty of Technology.

Acknowledgments: The authors wish to thank the two anonymous external reviewers that provided useful feedback on the basis of which the paper could be improved.

Conflicts of Interest: The authors declare that they have no conflicts of interest.

References

1. Global Geopark Network. Available online: http://www.globalgeopark.org/ (accessed on 21 June 2018).
2. United Nations Educational, Scientific and Cultural Organisation. Convention Concerning the Protection of the World Cultural and Natural Heritage. Available online: https://whc.unesco.org/en/conventiontext/ (accessed on 24 August 2018).
3. European Commission. Safeguarding cultural heritage from natural and man-made disasters. In *A Comparative Analysis of Risk Management in the EU*; EU Publications: Luxembourg, 2018; pp. 54–57.
4. Brimblecombe, P. Refining climate change threats to heritage. *J. Inst. Conserv.* **2014**, *37*, 85–93. [CrossRef]
5. Lemieux, C.J.; Groulx, M.; Halpenny, E.; Stager, H.; Dawson, J.; Stewart, E.J.; Hvenegaard, G.T. "The End of the Ice Age?": Disappearing World Heritage and the Climate Change Communication Imperative. *Environ. Commun.* **2017**, *12*, 1–19. [CrossRef]

6. Forino, G.; MacKee, J.; von Meding, J. A proposed assessment index for climate change-related risk for cultural heritage protection in Newcastle (Australia). *Int. J. Disaster Risk Reduct.* **2016**, *19*, 235–248. [CrossRef]

7. Phillips, H. Adaptation to climate change at UK World Heritage sites: Progress and challenges. *Hist. Environ. Policy Pract.* **2014**, *5*, 288–299. [CrossRef]

8. Phillips, H. The capacity to adapt to climate change at heritage sites—The development of a conceptual framework. *Environ. Sci. Policy* **2015**, *47*, 118–125. [CrossRef]

9. Kaslegard, A. *Climate Change and Cultural Heritage in the Nordic Countries*; TemaNord 2010:599; Nordic Council of Ministers: Copenhagen, Denmark, 2011; pp. 9–18.

10. La Russa, M.F.; Belfiore, C.M.; Comite, V.; Barca, D.; Bonazza, A.; Ruffolo, S.A.; Crisci, G.M.; Pezzino, A. Geochemical study of black crusts as a diagnostic tool in cultural heritage. *Appl. Phys. A* **2013**, *113*, 1151–1162. [CrossRef]

11. Rajčić, V.; Skender, A.; Damjanović, D. An innovative methodology of assessing the climate change impact on cultural heritage. *Int. J. Archit. Herit.* **2018**, *12*, 21–35. [CrossRef]

12. Jacob, D. A note to the simulation of the annual and inter-annual variability of the water budget over the Baltic Sea drainage basin. *Meteorol. Atmos. Phys.* **2001**, *77*, 61–73. [CrossRef]

13. International Strategy for Disaster Reduction. 2018 Summary of the HYOGO Framework for Action 2005–2015. Available online: https://www.unisdr.org/files/8720_summaryHFP20052015.pdf (accessed on 26 June 2018).

14. Romão, X.; Paupério, E.; Pereira, N. A framework for the simplified risk analysis of cultural heritage assets. *J. Cult. Herit.* **2016**, *20*, 696–708. [CrossRef]

15. Hambrecht, G.; Rockman, M. Approaches to Climate Change and Cultural Heritage. *Am. Antiquity* **2017**, *82*, 627–641. [CrossRef]

16. Goudie, A.S. Quantification of rock control in geomorphology. *Earth Sci. Rev.* **2016**, *159*, 374–387. [CrossRef]

17. Grönholm, S.; Alviola, R.; Kinnunen, K.A.; Kojonen, K.; Kärkkäinen, N.; Mäkitie, H. *Retkeilijan Kiviopas*; Geological Survey of Finland/Geologian Tutkimuskeskus: Espoo, Finland, 2006. (In Finnish)

18. Gomez-Heras, M.; McCabe, S. Weathering of stone-built heritage: A lens through which to read the Anthropocene. *Anthropocene* **2015**, *11*, 1–13. [CrossRef]

19. Wager, H. *Päijät-Hämeen Rakennettu Kulttuuriympäristö*; Päijät-Häme Regional Council Publication: Lahti, Finland, 2008. (In Finnish)

20. National Land Survey of Finland. MapSite. Available online: https://asiointi.maanmittauslaitos.fi/karttapaikka/?lang=en (accessed on 27 June 2018).

21. IPCC. Chapter 9: Evaluation of Climate Models. In *Climate Change 2013: The Physical Science Basis*; Cambridge University Press: New York, NY, USA, 2013; Available online: http://www.ipcc.ch/pdf/assessment-report/ar5/wg1/WG1AR5_Chapter09_FINAL.pdf (accessed on 24 August 2018).

22. Riahi, K.; Rao, S.; Krey, V.; Cho, C.; Chirkov, V.; Fischer, G.; Kindermann, G.; Nakicenovic, N.; Rafaj, P. RCP 8.5—A scenario of comparatively high greenhouse gas emissions. *Clim. Chang.* **2011**, *109*, 33. [CrossRef]

23. Daly, C. The design of a legacy indicator tool for measuring climate change related impacts on built heritage. *Herit. Sci.* **2016**, *4*, 19. [CrossRef]

geosciences

MDPI

Article

A Methodology for Long-Term Monitoring of Climate Change Impacts on Historic Buildings

Annika Haugen [1,*], Chiara Bertolin [2], Gustaf Leijonhufvud [3], Tone Olstad [1] and Tor Broström [3]

1   Norwegian Institute for Cultural Heritage Research, Storgata 2, 0155 Oslo, Norway; tone.olstad@niku.no
2   Faculty of Architecture and Design, Department of Architecture and Technology, Norwegian University of Science and Technology, Alfred Getz vei 3, 7491 Trondheim, Norway; chiara.bertolin@ntnu.no
3   Department of Art History, Uppsala University, 752 36 Uppsala, Sweden; gustaf.leijonhufvud@konstvet.uu.se (G.L.); tor.brostrom@konstvet.uu.se (T.B.)
*   Correspondence: annika.haugen@niku.no; Tel.: +47-41643690

Received: 9 July 2018; Accepted: 27 September 2018; Published: 4 October 2018

Abstract: A new methodology for long-term monitoring of climate change impacts on historic buildings and interiors has been developed. This paper proposes a generic framework for how monitoring programs can be developed and describes the planning and arrangement of a Norwegian monitoring campaign. The methodology aims to make it possible to establish a data-driven decision making process based on monitored decay related to climate change. This monitoring campaign includes 45 medieval buildings distributed over the entirety of Norway. Thirty-five of these buildings are dated to before 1537 and include wooden buildings as well as 10 medieval churches built in stone while the remaining 10 buildings are situated in the World Heritage sites of Bryggen, in Bergen on the west coast of Norway, and in Røros, which is a mining town in the inland of the country. The monitoring is planned to run for 30 to 50 years. It includes a zero-level registration and an interval-based registration system focused on relevant indicators, which will make it possible to register climate change-induced decay at an early stage.

Keywords: climate change; long-term monitoring; Norwegian protected buildings; medieval buildings; zero status; warning report

1. Introduction

The impact of climate change on a built cultural heritage must be considered in the long-term management of historic buildings. Most buildings are not constructed to resist "new" climate conditions and risks associated with climate change should, therefore, be identified and quantified in order to facilitate relevant adaptation measures.

A long-term strategy to adapt to climate change must be based on risk assessment, adaptation measures, and monitoring. Adaptation defines any adjustments in a system in response to actual or projected climatic stimuli [1,2] including changes in socio-environmental processes, practices, and actions to reduce potential damages.

Advanced simulations can be used to predict the future climate and its effect on historic buildings [3,4]. However, due to the high degree of uncertainty in simulations, this approach is not sufficient. Long-term monitoring of actual climate change impact is necessary to better understand the effects of climate change on historic buildings. Monitoring can be used to observe and analyze decay progress and changes to make them "visible" and to provide reference data to improve the results of simulation models. Additionally, monitoring can be used to inform decisions on adaptation actions and/or corrective actions. Furthermore, the outcome of monitoring provides results that can be communicated in order to raise awareness among property owners, heritage managers, and citizens and to gain political and economic support locally, regionally, and nationally.

Effective monitoring methods for the built heritage need to be identified and developed and indicators were specified. A good monitoring method for a built heritage may also be applicable to cultural heritage in general.

Climate model projections for Northern Europe indicate trends of increasing temperature and relative humidity and of extreme weather events such as floods triggered by local intense precipitation events and windstorms leading to direct and indirect damages on the built environment. Direct effects of climate change are manifested as physical changes to building structures and finishes. These can further increase once the actual local climate ceases to match the past environmental conditions that inspired the building design (SMOOHS Project, www.smoohs.eu, Predicting and managing the effects of climate change on World Heritage). Both direct and indirect effects may affect the value of the built environment, which further drove the loss of important components of individual and collective identity of a historic site.

The impacts of climate change may be greatly amplified on the aged and fragile materials that are present in historic buildings. Advanced adaptation techniques, which are commonly used for modern buildings and structures, cannot always be applied due to legal requirements of preserving the original features of historic buildings.

The need to use traditional materials, construction systems, and craft skills during maintenance and refurbishment interventions is already well recognized in Norway. We suggest that there is now an urgent need to evaluate successes and failures of such interventions in the face of climate change. Long-term monitoring can provide essential insights into how traditional materials and construction systems might be modified to cope with more aggressive climate conditions.

Despite the growing interest in understanding the effects of climate change on cultural heritage, there have been few attempts to assess actual impacts through long-term monitoring [5]. According to a recent review by Fatorić and Seekamp 2017 [6], there has also been a lack of studies focusing on the implementation and documentation of adaptation measures taken to protect cultural heritage from climate change. Long-term monitoring is key to better understanding both impacts and the need for adaptation measures. Moreover, monitoring is essential for determining the causality between climate and damage at individual sites [7]. Long-term site-based monitoring, therefore, elucidates necessary feedback loops for the evaluation of adaptation planning and for the calibration of impact risk assessments.

An additional challenge for the monitoring of impacts of climate change is the long time frame needed to discern the climate change signal from the natural variability of the climate. This time frame (>30 years) exceeds the scope of most research funding schemes and there are also practical difficulties in maintaining such long projects with regard to administration, staff continuity, data retrieval and storage, etc. [5].

The objective of this paper is to present a novel methodology for monitoring long-term effects of climate-induced degradation on historic buildings and interiors. The first part of the paper presents a generic framework for the development of monitoring programs. This framework is based on a review of existing approaches to climate change monitoring of cultural heritage as well as the experiences from the Norwegian project known as "Methods for Monitoring the Effects and Consequences of Climate-Related Degradation of Buildings" (Metoder for overvåking av effekter og konsekvenser av klimabelastninger på bygninger, Forprosjekt, NIKU-rapport 197/2016). The second part of the paper presents, as a case study, the implementation of a newly started long-term monitoring campaign on the impacts of climate change on historic buildings in Norway.

State-Of-The-Art

Comprehensive approaches that consider the connection between structural and environmental aspects in the built environment are used to develop frameworks to continuously monitor the operational performance of buildings. These commonly concentrate on the energetic aspect of modern buildings in operation. Within such frameworks, real-time monitoring is proposed to collect data about building performance with the final objective being to better control operations within the buildings, which ensures a reduction of building energy consumption.

The impacts of future climate change on cultural heritage is an expanding research field where the majority of studies over the last 10 years have dealt with climate change impact assessment and adaptation planning [6]. A number of European studies have performed regional risk assessment (RRA), which is a methodology to appraise the risk posed mainly by hydrological-related climate change impact [8,9], but also by a range of mechanical, chemical, and biological damage on a set of generic building types and mixed materials [3]. The methodology, in the case of risks related to the hydrological cycle (e.g., storm surge, floods, sea level rise, moisture, rainfall, and coastal erosion), is based on the concept that risk is a function of hazard, exposure, and vulnerability. It integrates the output of various hydrodynamic and climatic models with site specific geophysical and socio-economic indicators (e.g., key performance indicators as in Gandini et al., 2017 [10]) to develop risk indices and GIS-based maps [9]. The focus has often been on wide areas where multiple objects are located, i.e., recent and new buildings, historical buildings, monuments, infrastructure, and landscape. Over the last 10 years, the projects dealing with this approach have included ADVICE and KULTURisk. Their objectives were to improve knowledge of vulnerability by using, for example, the LIDAR (Laser Imaging, Detection and Ranging) technique to obtain a high-resolution coastal topography on exposure by using a wide range of key performance indicators on hazards by improving hydro-climatic models and on developing a culture of risk prevention by proposing structural and non-structural mitigation measures. The method, in the case of outdoor and indoor multiple risks, is based on dose-response and damage functions. It integrates the output of climatic models using moderate emission scenarios (considering socio-economic factors that may influence the magnitude of climate variables) with a set of generic building and material specific indicators [3,11,12] to develop risk indices as well as risk and decay maps. Its focus is on a single object category, i.e., cultural heritage, with the aim of improving knowledge of cumulative and slow climate-induced mechanisms of decay on a wide range of materials and of building simulation. Specifically, the building simulation is based on monitored data from existing buildings used to effectively transfer outdoor climatic conditions to indoor conditions.

A recent review [13] showed the variety of ways that climate change might have an impact on the cultural heritage. The review summarizes how different hazards that are intensified by climate change (temperature change, precipitation change, storm surge, flood frequencies, coastal erosion, sea level rise, carbon dioxide levels, and combined stressors) can have an impact on different cultural resources. This review is useful for identifying and selecting which phenomena to monitor in long-term monitoring campaigns of climate change impacts.

Over the last 10 years, two major European projects have attempted to assess the future climate change effect on tangible cultural heritage including The Noah's Ark project, which focuses on outdoor risks, and the Climate for Culture project, which deals with both indoor and outdoor risks.

A few other European studies have assessed the impacts of climate change on buildings and interiors by combining the projections of future climate change with damage functions [4] and, more recently, in combination with hygrothermal building simulations [3,14–16]. These top-down impact studies, which are mainly focused on predictions rather than monitoring, show that assessments of climate change impacts on historic buildings and interiors are both complex and uncertain. Nevertheless, taken together, these studies show that climate change will cause significant damage to historic buildings and historic interiors in Europe unless plausible adaptation measures are implemented.

Based on a literature review and three national adaptation strategies, Klostermann et al., 2018 [17] presented a generic strategy for monitoring based on four key factors.

(1)   definition of the system of interest,
(2)   selection of a set of indicators,
(3)   identification of the organizations responsible for monitoring,
(4)   definition of monitoring and evaluation procedures.

Although this review mainly concerns the monitoring of climate change adaptation, it provides a generic approach that we suggest is also applicable to the monitoring of climate change impact. Table 1 shows how each of the four key factors are specified by a number of questions. In Table 2, general challenges are presented as well as general proposed solutions.

Table 1. Questions for the design of a monitoring program from Klostermann et al. (2018).

	Questions
System of interest	1. Is the description of the monitoring context based on a transparent and structured overview of:
	a. Current and future climate (preferably on the basis of downscaled climate models)?
	b. Important climate impacts on socio-economic and environmental systems including exposure and sensitivity?
	c. Socio-economic and environmental vulnerabilities?
Indicators	2. What indicators are selected for monitoring and evaluation?
Responsible organization	3. Which organization(s) is/are responsible for monitoring?
	4. What financial and other resources are available to the organization for monitoring?
	5. What are the arrangements that provide legitimacy and credibility for the monitoring?
Procedures	6. Are information needs and monitoring objectives clearly described?
	7. Are monitoring procedures clearly specified including data collection and reporting?
	8. Do the procedures prescribe stakeholder involvement and, if so, where in the monitoring process?
	9. Is the notion of adaptive monitoring incorporated?

Table 2. General challenges and solutions for monitoring. Adapted from Klostermann et al. (2018).

General Challenges for Monitoring	Proposed Solutions
Useful information: salient and context sensitive to specific information demands.	• Involve stakeholders to check information needs. • Research mechanisms in system(s) of interest.
Technical quality of indicators: accurate, valid, precise, robust, meet SMART criteria. (SMART stands for: Specific, Measurable, Assignable, Realistic, Time related)	• Use/develop review procedures. • Use existing indicators/data sources. • Research physical mechanisms in system(s) of interest.
Communicative value and efficiency of indicators: simple and straightforward to understand.	• Test communicative value of indicators. • Use existing well-known indicators.
Credible production of information: unbiased, legitimate, transparent, objective/independent.	• Scientifically sound methods. • Independent operation of the monitoring organization.
Monitoring must be feasible: availability of data and limited financial and human resources.	• Limit the set of indicators. • Use existing datasets. • Evaluate usefulness of indicators.

There are some challenges to overcome when monitoring climate change impacts on cultural heritage related to the time scales involved and to the connected problem of data retention. The need for improved monitoring is often emphasized in reports and guidelines that are produced by agencies and organizations in charge of cultural heritage and is aimed at decision makers and stakeholders without giving much detail about how such monitoring should be carried out in practice (e.g., National Park Service 2010 [18], English Heritage 2006 [19]).

To our knowledge, the only research project that has focused solely on the long-term monitoring of climate change impacts on cultural heritage is the PhD project of Cathy Daly [5,7]. Daly developed a methodology and a tool for monitoring the impacts of climate change on archaeological sites. The methodology consisted of a vulnerability and impact assessment framework to be used for each site and the suggested monitoring tool is a sacrificial stone object, which tracks surface changes caused by recession, salt crystallization, and microbiological growth. The monitoring tool is supposed not to require maintenance. The most important contribution from Daly is that she identified and analyzed the needs of a robust method that is supposed to be applied in practice. Her study clearly demonstrates that cultural heritage managers cannot rely on existing approaches to monitoring and that there is a need to think in new ways about how monitoring should be carried out.

An overview of a recent project to monitor climate change impacts on cultural heritage in Egypt and the UK is presented in Mahdjoubi et al., 2017 [20]. It is unclear how the presented methodology was developed, but, nevertheless, it argues for some innovative ways of monitoring the long-term degradation of heritage buildings with the use of laser scanning, photogrammetry, air permeability measurements, and U-value measurements.

The state-of-the-art approaches reported above highlight an urgent need for the heritage sector to initiate long-term monitoring campaigns about the impacts of climate change to historic buildings.

The review also shows that, while there is abundant knowledge about theoretical cause–effect relationships between microclimates and building damage, there is a lack of knowledge and experience about how to set up and manage long-term monitoring projects and to plan and implement a data driven decision-making and adaptation process.

A data-driven decision-making method focused on the monitoring of climate-induced decay in valuable buildings will have to take into account the fact that the historic value of historic buildings restrict the range of possible and allowable interventions. Conservation interventions cannot be standardized. They need to be tailored, non-invasive, effective, and sustainable in term of costs and resource use. Additionally, a team with interdisciplinary knowledge and expertise is required since the understanding of the monitoring results and the conservation needs are based on an integrated and comprehensive assessment of climate change impact. Several factors have to be considered when interpreting the results of monitoring at individual sites: the building structure and constituting materials, the possible existence of induced natural decay, the decay rate assessment, the history of past maintenance intervention, and the expected or predicted building lifetime.

2. Materials and Methods

2.1. Development of a Method for the Long-Term Monitoring of Climate Change Impacts on Cultural Heritage

2.1.1. A Generic Framework for Developing Monitoring Programs

In 2016, the Directorate for Cultural Heritage in Norway asked The Norwegian Institute for Cultural Heritage Research (NIKU) to develop a program for the long-term monitoring of climate change impacts to historic buildings. In order to develop the program, a multi-disciplinary team of experts was gathered and the work proceeded in a systematic way, which is presented in Figure 1. Based on our experiences, we suggest that this generic approach can be a valuable framework for the development of monitoring programs in other contexts (Figure 1).

The first step is to define and understand the system of interest or the target of the monitoring by:

- Understanding the historic building through the collection of data on location, type of immovable cultural heritage, type of material and state of preservation, and statement of significance, authenticity, and integrity to understand aspects to be safeguarded.
- Understanding the hazards, i.e., dangers and threats to the area including those induced by climate change, use of the building, existing policies, strategies, plans, and actions that are of relevance for preventive conservation and maintenance.
- Understanding the instruments of safeguarding, i.e., national law, legislation, local regulations, international conventions/charters, and, later in the project, to understand if these instruments have to be adapted or new ones have to be developed.

The second step is to gather a multidisciplinary team with specialized competence in building physics, monitoring, conservation, climate, different kinds of damage (biological, chemical, and mechanical), and climate change impact monitoring (Figure 1, second column, step 2). The combined skills of the group should be tailored for the system of interest, i.e., the target of the monitoring. By establishing a project team with the necessary competence, it is possible to quickly develop and implement the monitoring project, perform the analysis of the collected data, identify the effect of climate change on the historic fabric, and also to find the parts of the buildings that are most vulnerable to climate-related decay. The interdisciplinary expert team should be carefully chosen and, to ensure competence and continuity, it should preferably include younger members. Additionally, a second group of relevant stakeholders is involved at an early stage. This is composed of relevant public and private stakeholders that can take actions once monitoring reporting has been delivered. During this step, organizational, operational structures and procedures for further developing actions to safeguard and adapt historic buildings to climate change effects (e.g., through ordinary and extraordinary maintenance) have to be decided.

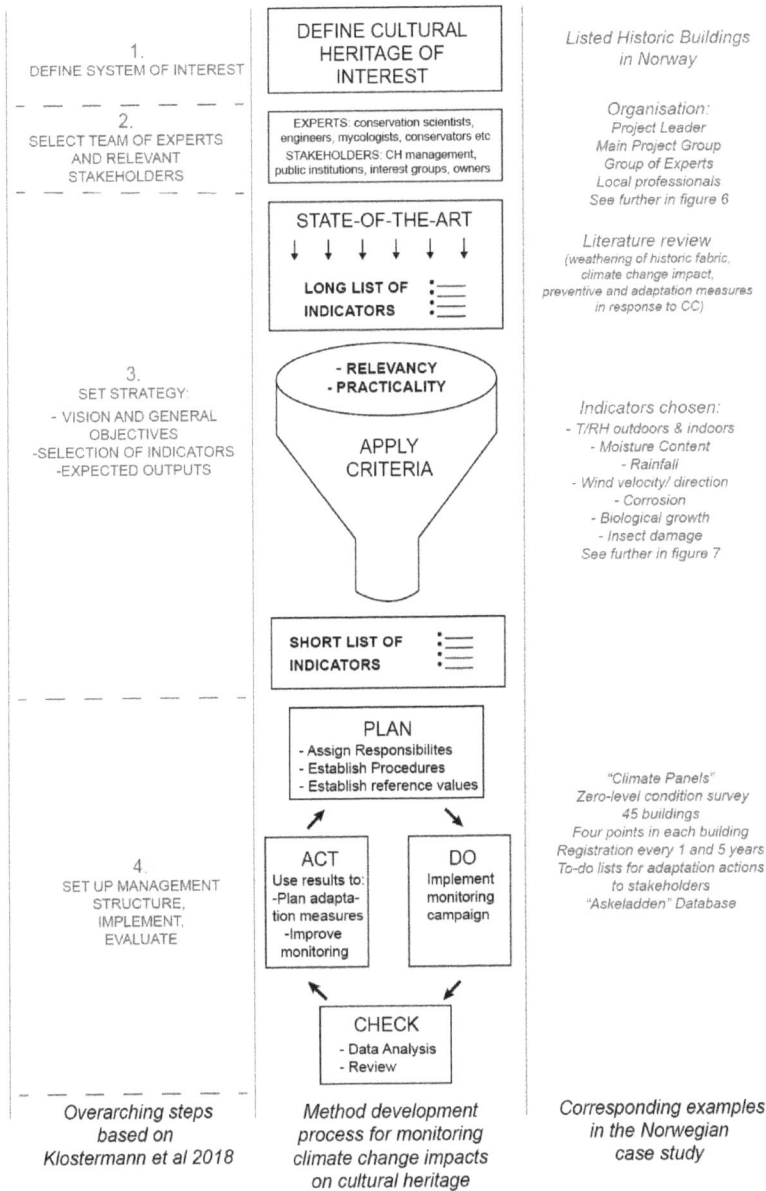

| Overarching steps based on Klostermann et al 2018 | Method development process for monitoring climate change impacts on cultural heritage | Corresponding examples in the Norwegian case study |

Figure 1. A generic framework for developing monitoring programs targeting specific cultural heritage objects. The first column shows the overarching steps (adapted from Klostermann et al., 2018), the second column shows the method development process for monitoring climate change impacts on cultural heritage, and the third column shows corresponding examples in the Norwegian case study.

The third step (Figure 1, second column, step 3) involves deciding the strategy and includes:

- The overall vision and general objectives of the monitoring: (1) understanding the field of actions (i.e., adaptation of cultural heritage to climate change effects by minimizing their negative impacts); (2) acquaintance with the state-of-the-art, i.e., identification of relevant issues to deal with (e.g., weathering of historic fabric, preventive and adaptation measures in response to climate change), and, consequently; (3) objectives to achieve and strategies (integrated approach).

- The understanding of how to achieve the goals, i.e., identifying the environmental variables, actions, instruments, and techniques (both existing and to be developed with a main focus on a non-destructive technique (NDT)) to accomplish the objectives of the long-term monitoring. These are the outputs achieved as the recognition of alterations ascribed to climate change when compared to the zero status registration.

Lessons learned from other fields such as agriculture and human health help to determine how to reach the objectives, e.g., the selection of domain-specific indicators for climate change impact. Clear objectives help to discern intermediate steps and/or important selection criteria for reaching the ultimate goal. Related to the objectives of a long-term monitoring program, it is important to distinguish between measurements and indicators: Measurements concern changes in physical, quantitative attributes while indicators are used to describe or project the performance of a system in a quantitative or qualitative way [21]. The initial step is to find a systematic method with which to identify useful indicators among those available in the literature and to include those indicators in a final monitoring plan. During our discussions in the project group, two selection criteria emerged as being the most important, which are relevancy and practical relevancy. For an indicator to be relevant for monitoring the impacts of climate change, it must be possible to determine if a change was caused by climate change or other factors. The indicator has to address specific and overall vulnerability and exposure related to the effects of climate change based on a starting condition level, i.e., a zero status registration. If the cause–effect relationship is too uncertain, there will be little or no possibility of discerning which changes were actually caused by climate change and which were caused by other factors. The involved uncertainties can be related to measurement errors (input to the model), deficiencies of the model itself, and the natural variability of the cause–effect relationship. Dose–response relationships are the easiest to model. However, climate-induced damage is often more complex involving synergisms and, hence, is more difficult to model [22]. Practicality. Given the time frame, available resources (financial and competence) and the availability of existing technologies, it is necessary to prioritize indicators that are possible to monitor in practice. Non-destructive techniques (NDT) are preferred for the detection of environmental conditions so as to avoid damage to the building structure. However, very few such techniques are routinely used in cultural heritage conservation as a number of barriers that are both technical and institutional and hinder their in situ implementation on a long-term basis.

The fourth step (Figure 1, second column, step 4) is setting up the management structure of the monitoring program. The proposed strategy is based on the "plan-do-check-act" cycle that employs long-term monitoring and its review to take adaptation actions in time. The four phases of this cycle have to meet the following requirements:

- *Plan*: Development and/or adaptation of organizational and operational structures and procedures clearly allocating responsibilities and tasks to be met (e.g., through an organigram that explains a responsible body/person and sets the individual goals for: (1) safeguarding the cultural heritage values, (2) the implementation, application, and revision of monitoring and data collection, and (3) assuring that actions are taken in time, according to monitoring results.
- *Do*: Implementation plan and execution phase of long-term monitoring. The implementation plan includes a sequence of actions (i.e., tasks and activities as systems for collection and storage of data, reporting and management, etc.), which have to be executed (e.g., who is doing what, what inputs are needed, what outputs are intended, etc.). Additionally, the collection of information concerning the assessment of the building or structure to define the so-called zero-level registration condition (step 1) has to be concluded before starting the monitoring (execution phase). This defined state will be the base for the future assessments and registrations of the object that is to be monitored. In this phase, a close communication with the institution that is managing the historic buildings is needed.

- *Check*: Continuous monitoring and review (e.g., description of who is monitoring or evaluating what and how results will be used). This checking or review stage will strengthen the evidence base in time for the next cycle of reporting so that the historic building will be better served by the method/process itself.
- *Act*: Acting to adapt historic buildings to climate change effects on the basis of monitoring results. In the long-term perspective, the program will evaluate adaptive measures that have been proposed and eventually implemented by directly or indirectly monitoring them.

2.1.2. Suggested Monitoring Method

The suggested monitoring method is based on a zero-level condition survey (first step) focusing on possible climate-induced degradation processes, which is followed by a control survey every fifth year (fourth step and iteration of first step) depending on the category of the building. The condition survey is based on the European standard NS-EN 16096:2012 [23]—Conservation of cultural property—condition survey and report of built cultural heritage. The zero-level condition survey starts with the identification of the (1) building involving the site description, climate conditions, environment, construction, materials, age, and condition including the history of damages and the history of maintenance; (2) predicted future change involving temperature, relative humidity, solar radiation, wind, and precipitation; and the (3) risk for the climate involving induced disasters on the site such as floods, avalanches, and rockslides. The focus of the monitoring is the slow influence of the climate on the monitored cultural heritage.

Further condition surveys and documentation during the implementation of the long-term monitoring (fourth step) are dominated by climate logging combined with visual control and detailed photos. Climate logging is performed by the use of "climate panels."

2.1.2.1. The Climate Panels

"Climate panels" provide a standardized continuous record (registration every hour) of temperature (T), relative humidity (RH), and moisture content (MC) and information on biological risk. A climate panel mounts specimens or blocks and meters on a non-organic, lightweight sheet material. The following are mounted: (1) a T and RH logger (Materialfox from Scanntronik Mugrauer GmbH selected for the Norwegian monitoring program); (2) an MC logger (i.e., an electrical resistivity probe inserted into a specimen of the same material that the meter is calibrated for); and (3) a certain number of standardized specimens or small blocks used as test material to provide an early warning about conditions that favor the proliferation of microorganisms. All climate panels must have a set of standard blocks of a single type of material but may also have additional materials adapted to issues specific to the location.

Climate panels should be placed in carefully chosen parts of the building. It is possible to perform analyses of salts, bacteria, algae, and the identification of insects on the specimens. The specimens are evaluated each time a subsequent registration is made in the building through visual inspection by recording visible growth or other visual changes in the control material. Photography and sampling and analysis of the surface layer of the block closest to the logger are performed at the first follow-up and then are repeated for block 2 from the logger for the second follow-up, and more. Indirect measurements of the moisture ratio in adjacent blocks can be made through comparison between climate panels. Monitoring using climate panels makes it possible to define the damage from and symptoms of the moisture impact on building materials (e.g., progress of damage, new damage, and symptoms) in relation to the zero status registration mentioned above.

The climate panels should be placed approximately in the same positions in all the monitored buildings in order for the results to be comparable. An example of standardized placement is shown in Figures 2 and 3.

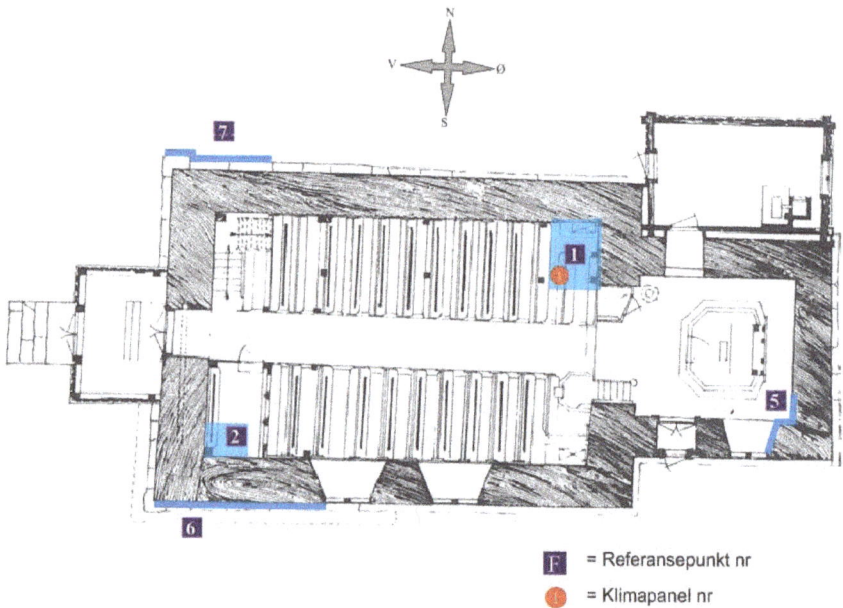

Figure 2. Positions of climate panels (red dot) and reference points (blue squares) in Skoger Old Church.

Figure 3. Climate panel. Photo: NIKU.

In the present study, four fixed reference points are chosen in each building and the climate panels are placed in two of these. These four fixed reference points are situated in the parts that are likely to be most and least threatened by moisture, which are the northeast and southwest corners, respectively, both on the ground level and at the highest levels in the buildings. The climate panels are placed in the northeast corners, which are likely to have the highest levels of moisture and, thereby, are also likely to be the most threatened.

2.1.2.2. The Reports

The results of the on-site surveys as well as the collected information from the logging equipment are presented as processed information in reports for each monitored building.

There is one report for the zero status registration (first step) and reports for each of the following control registrations (check phase in the fourth step). The expert group will perform the analysis and prepare the reports. The degree to which observed alterations at any of the selected reference points for each building or observations of a general change of the situation at the site are of a nature that will increase the risk for the building in the future climate will be reported. If needed, measures and further examinations will be proposed in a report (e.g., changes in the monitoring plan and implementation following the notion of adaptive monitoring and/or proposal of adaptation measures to apply on the building).

Additionally, if a set value in the monitoring is exceeded and indicates increased degradation between two surveys, a warning must be generated from the onsite local person involved in the project. A warning report will be produced. The observed deviation in the monitoring may be difficult to define and the onsite local person must have the ability to involve experts immediately if there is any uncertainty. The report will describe the problem and propose necessary measures or a possible change in the monitoring focus for the site in question. If the situation is critical, the expert team should be notified and a summary of the most important information with the raw data enclosed will be included in the warning report. The project's expert team is responsible for handling these situations and for the elaboration of the warning reports. If actions are needed, the manager of the building has the responsibility for implementing measures.

A huge amount of data is collected during a monitoring program. It is necessary to collect enough high-quality data for the aim while simultaneously obtaining a minimum of data. The challenge is to define exactly what data is needed for the monitoring. It is a demanding task to both find systems that, with necessary modifications, will run for a long period of time and find a way to make it possible to compare data for the duration of the monitoring period. Comprehensive reports for each building must be made with certain intervals during this long period. These may be used for comparison in case the access to the raw data becomes difficult or impossible.

It is necessary to use an official database that will be maintained in the future and which contains information on cultural heritage buildings. A system for storing the information and photos must be developed in cooperation with the owner of the database.

In principle, the aim of the reports and storage of data is to have a transparent system with a high grade of accessibility for as many data items as possible. The baseline requirement is that all levels of management should have access to the raw data. The owner of the project will own the information database and handle the formalities related to the system of accessibility and the system of transferring data to the managers and researchers.

3. Results

3.1. Implementation of the Method in a Monitoring Program in Norway and the First Results

3.1.1. Selection of Buildings

The selection of buildings was based on a set of criteria outlined in Figure 4. These criteria might be used on a general basis as part of the monitoring model.

Based on the abovementioned criteria, 45 buildings were chosen to be included in this study. As demanded by the Directorate for Cultural Heritage, 35 of these buildings dated to before 1537 and 10 of these buildings were medieval churches built in stone. The remaining 10 buildings are situated in the World Heritage sites of Bryggen in Bergen on the west coast of Norway and in Røros, which is an inland mining town east of Trondheim. The monitoring is planned to run for 30 to 50 years.

The selected buildings are geographically distributed over the country in various climatic zones and in different landscape sites and situations within the zones such as in valleys, mountainous areas, forests, or close to water. Buildings close to a weather station and with rainfall and/or temperature data already monitored were preferred.

The risk of damage was considered as the most important criterion and the exterior was valued as more important than the interior in terms of authenticity and original or old building elements. We avoided excessively complex buildings. The type, function, and use of buildings were determined to be less important criteria since the old buildings were mainly of the same type. Distance to other selected buildings was considered to be of high importance for the economy and the implementation of the monitoring project. Availability signifies accessibility to the building including an accommodating owner.

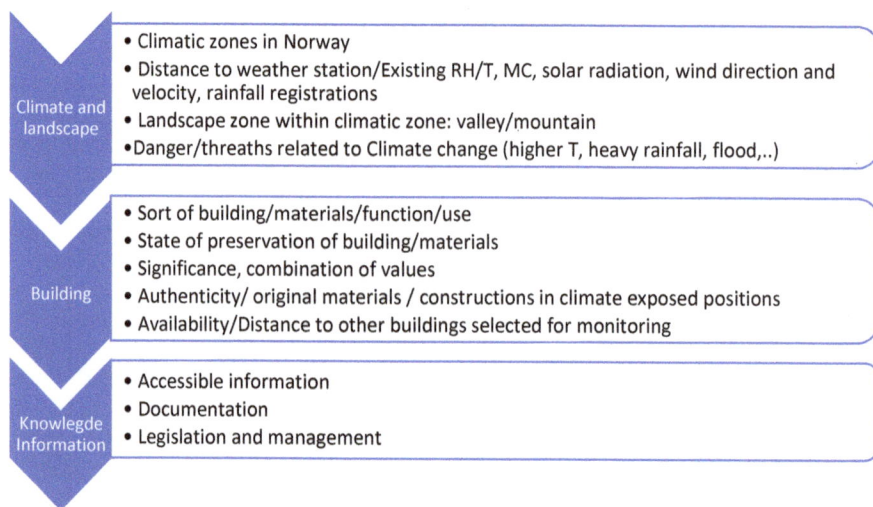

Climate and landscape
- Climatic zones in Norway
- Distance to weather station/Existing RH/T, MC, solar radiation, wind direction and velocity, rainfall registrations
- Landscape zone within climatic zone: valley/mountain
- Danger/threaths related to Climate change (higher T, heavy rainfall, flood,..)

Building
- Sort of building/materials/function/use
- State of preservation of building/materials
- Significance, combination of values
- Authenticity/ original materials / constructions in climate exposed positions
- Availability/Distance to other buildings selected for monitoring

Knowlegde Information
- Accessible information
- Documentation
- Legislation and management

Figure 4. Criteria for the selection of buildings. RH = Relative Humidity, T = Temperature, and MC = Moisture Content.

Documented buildings and buildings that were well known to the project team were preferred.

We found that availability of buildings was crucial. Additionally, to secure a long-period contract for the monitoring, only cultural heritage sites owned or managed by the public should be selected for monitoring to ensure that they are available for the whole monitoring period.

The monitoring project was started in July 2017 with the following four buildings:

- Skoger Old Church, which is a stone church dating to around 1200 and is located in the inland region near Oslo.
- Raulandstua, which is a timber building dating to 1238, which has been hosted at the Oslo open-air museum since 1899.
- Garmo stave church, see Figure 5, which is a wooden building originally dating to early 1200 and is now hosted at the Maihaugen museum, Lillehammer, in the inland region north of Oslo. It was erected at the museum in 1921.
- Bugarden, which is a wooden 18th century building at the World Heritage Site of Bryggen (the Wharf) in Bergen on the west coast.

3.1.2. Organization

The organization includes both the expert team and onsite local persons, see Figure 6. The expert team started and implemented the monitoring and established the close and necessary communication with local persons. A control of the registration from the climate panel made after one year and the following control registrations (every fifth year) were made by the interdisciplinary team of experts.

Figure 5. Garmo Stave Church, Lillehammer. Photo: NIKU.

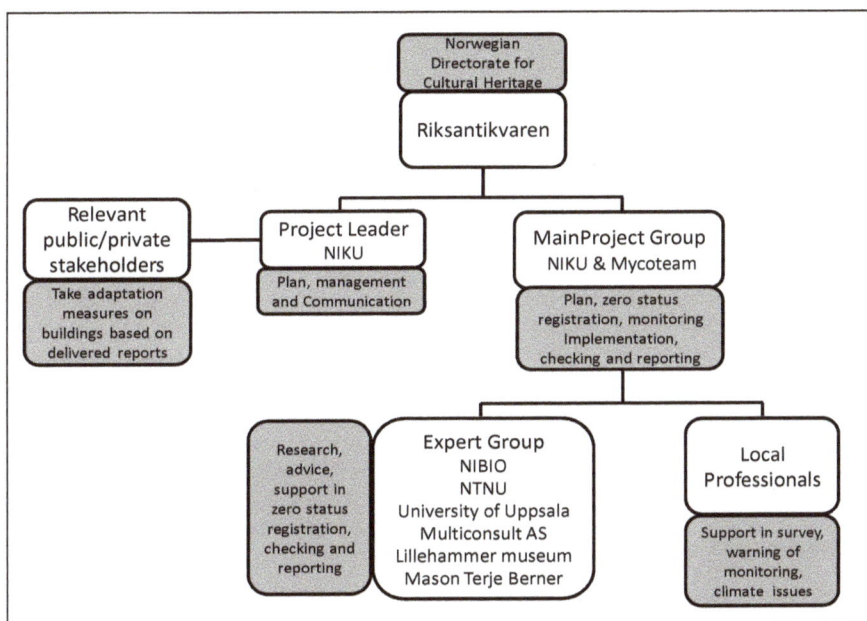

Figure 6. Organigram of the monitoring program: Environmental monitoring of the impact of climate change on protected buildings—Miljøovervåkning av konsekvensene av klimaendringene på fredete bygninger (2017–2026). Project Number SAKSNR 15/02185 funded by Riksantikvaren—Norwegian Directorate for Cultural Heritage. White Boxes: Responsible body/person. Grey Boxes: Set of tasks.

3.1.3. Choice of Indicators

Selected indicators used in the monitoring program are outdoor and indoor climate parameters (e.g., T and RH), weather variables (e.g., precipitation, wind direction, and wind velocity), moisture-related parameters in building materials (e.g., MC), visible damage on wood and masonry such as cracks and voids, biological decay of wood and surfaces such as the growth of micro-organisms, observation of insects, salt crystallization, frost damage in masonry, and the flaking of painted surfaces on wood and masonry (see Figure 7).

Sometimes the change of a single indicator against the zero-level registration cannot give a clear measure of the effect of climate change. However, using several indicators in combination can provide a better overview of the effect and the cause of changes. Although the proposed methodology provides measures and, therefore, quantitative information, it might be necessary to consider an indication of change and/or a new risk development to determine if an adaptation action has to be taken or not. The indicators presented in Figure 7 were chosen based on our monitoring objectives and available resources for monitoring. However, the proposed scheme can be easily adapted to changing needs and objectives.

Objective	Indicator	indicators and risk levels control against the zero level registration		Verification date	CC scenarios for comparison
		Warning code	Explanation		
Climate Change effect on historic buildings	T, RH, MC, rainfall,..., biological risk		Low (<<) Change of indicators and risk levels → No need of adaptation intervention(s)	Period: last 1,..,5 year(s)	https://www.climateforculture.eu/
			Medium Change of indicators and risk levels → need of adaptation intervention(s)	Last data collection and reporting: 01.08.2018	http://www.miljostatus.no/tema/klima/klimainorge/klimainorge-2100/
			High Change of indicators and risk levels → urgent and major adaptation interventions needed		

Figure 7. Setup of the indicators used in the long-term monitoring program to assess the impact of climate change on historic buildings.

3.1.4. The Zero-Level Condition Survey

For each building, former alteration and repair history were collected through an archival search, see Figure 8. Since damages often occur in the same parts of the buildings, the damage history is important. Studies of climate scenarios for the building sites and their surroundings were conducted as a background for the interdisciplinary zero-level condition survey (Figure 1, third column). A thorough survey was conducted in each building, which establishes the zero-level condition. During this survey, the most vulnerable parts of the buildings and their interiors were selected and reference points were established. The reference points include the four fixed points as well as possible supplementary points (see Section 2.1.2.1). Early warning reports for two of the buildings were immediately sent to the Directorate for Cultural Heritage

3.1.5. Information Storage and Access

It was decided together with the Directorate for Cultural Heritage that all information collected in the monitoring project would be added or linked to the official database of the Directorate for Cultural Heritage in Norway, which is named "Askeladden." This included: (1) The information from the "zero status survey" and photos for each building, (2) the raw data collected during the monitoring, and (3) the reports with the processed information. The system for storing the information in the database was developed in cooperation with the Directorate for Cultural Heritage. Once all the data has been stored in the database, the integration of a long-term monitoring program in a Cultural Heritage Integrated Management Plan (CHIMP) becomes possible through the implementation of the suggestion provided in the monitoring reports.

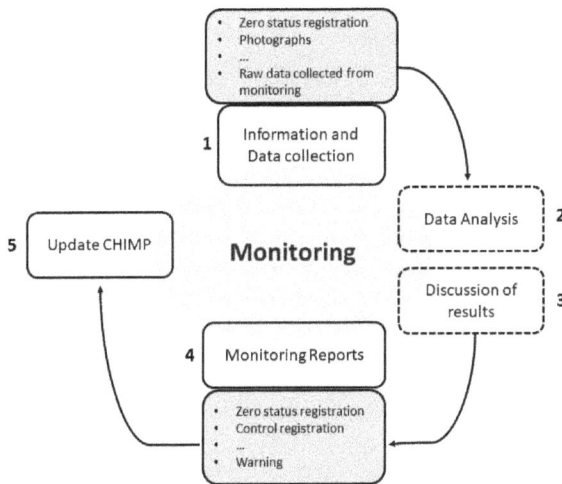

Figure 8. Incorporation of a long-term monitoring program in a Cultural Heritage Integrated Management Plan (CHIMP). Transparent boxes with solid border: information, data, and report publicly available in the official database of the Directorate for Cultural Heritage in Norway named "Askeladden." Boxes with dashed border: information available internally in the main project group. Gray boxes: examples of information, data, and reports within the monitoring program.

3.1.6. Future Plans for the Monitoring Project

In 2018, the project will include 10 additional buildings situated in Bergen in the northern part of the west coast and in Northern Norway. A thorough zero-level condition survey will be conducted in all of these buildings and the first year control will be performed on the four buildings that initiated the project in 2017.

Adding the experiences after half a year of monitoring, the expert group concluded that there is a need to focus on a solid systematic approach in the project. It has been recognized to be of great importance to document and mark photographs and log files in a well-established, systematic, and solid protocol to be able to correctly interpret results in the long term and to speed up corrective decisions concerning measures related to the monitoring.

4. Discussion and Conclusions

The development of an efficient and focused monitoring technique with results useful for the preventive conservation of cultural heritage depends, to a high degree, on a deep knowledge of the thresholds of decay that are acceptable or not acceptable to maintain the aesthetic, historic, and cultural values of aged and weathered materials. There is a need to define acceptable thresholds of several types of degradation processes through experience by conservators who work on real objects in situ as well as from experience gained during laboratory tests by heritage scientists working with imitation samples and with aging tests and/or simulations in climate chambers. However, during such laboratory work, it is important to involve to a high level an interdisciplinary reference group with competence in climate-related decay in buildings. Improved knowledge of degradation thresholds will allow conservators to intervene in time to mitigate the impacts of climate change on historic buildings by applying a preventive (early warning) intervention method.

More efficient monitoring could also be achieved by the use of refined and customized information technologies. For example, methods for uploading documents and monitoring data as well as by the systematic use of more detailed photos.

One of the main objectives of long-term monitoring is to calibrate and validate climate and building simulations aiming to assess future risks. In order to be able to use theoretically simulated climate conditions and degradation processes as a base for real monitoring in the future, it is necessary to further examine existing simulation programs, test them, and compare them with monitoring results. The results of the EU project Climate for Culture could be used for combining simulation results with existing monitoring results [24]. It will, thereby, be possible to both develop simulation models further and to establish more efficient monitoring programs. Additionally, by linking simulations and monitoring, it will be possible to better manage proactive restoration work. Through a closer connection between monitoring and adaptation, both risk assessments and the preservation of cultural heritage values will be improved, which is illustrated in Figure 9.

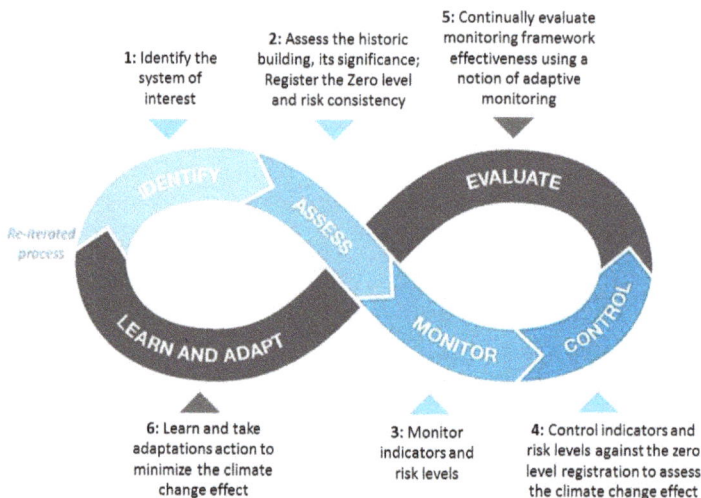

Figure 9. Through a closer connection between monitoring and adaptation, the risk assessment and preservation of cultural heritage values will improve.

Author Contributions: Conceptualization: A.H., C.B., G.L., and T.B. Methodology: all the authors. Investigation of the state-of-the-art: T.B., G.L., and C.B. Writing—original draft preparation: All the authors. Writing—review and editing: All the authors. Project administration: A.H.

Funding: We are grateful to the Riksantikvaren—Norwegian Directorate for Cultural Heritage for funding the project: Environmental monitoring of the impact of climate change on protected buildings—Miljøovervåkning av konsekvensene av klimaendringene på fredete bygninger (2017–2026). Project Number SAKSNR 15/02185.

Conflicts of Interest: The funders had no role in the design of the study, in the collection, analyses, or interpretation of data, in the writing of the manuscript, or in the decision to publish the results.

References

1. Klein, R.; Smith, J. Enhancing the capacity of developing countries to adapt to climate change: A policy relevant research agenda. In *Climate Change, Adaptive Capacity and Development*; Smith, J., Klein, R., Huq, S., Eds.; Imperial College Press: London, UK, 2003.
2. Smit, B.; Burton, I.; Klein, R.J.T.; Street, R. The science of adaptation: A framework for assessment. *Mitig. Adapt. Strat. Glob. Chang.* **1999**, *4*, 199–213. [CrossRef]
3. Leissner, J.; Kilian, R.; Kotova, L.; Jacob, D.; Mikolajewicz, U.; Broström, T.; Ashley Smith, J.; Schellen, H.; Martens, M.; van Schijndel, J.; et al. Climate for culture—Assessing the impact of climate change on the future indoor climate in historic buildings using simulations. *Herit. Sci.* **2015**, *3*, 38. [CrossRef]
4. Sabbioni, C.; Brimblecombe, P.; Cassar, M. *The Atlas of Climate Change Impact on European Cultural Heritage: Scientific Analysis and Management Strategies*; Anthem: London, UK, 2010.

5. Daly, C. A Cultural Heritage Management Methodology for Assessing the Vulnerabilities of Archaeological Sites to Predicted Climate Change, Focusing on Ireland's Two World Heritage Sites. Ph.D. Thesis, Dublin Institute of Technology, Dublin, Ireland, 2013.

6. Fatorić, S.; Seekamp, E. Are cultural heritage and resources threatened by climate change? A systematic literature review. *Clim. Chang.* **2017**, *142*, 227–254.

7. Cathy, D. The design of a legacy indicator tool for measuring climate change related impacts on built heritage. *Herit. Sci.* **2016**, *4*, 19.

8. Ronco, P.; Gallina, V.; Torresan, S.; Zabeo, A.; Semenzin, E.; Critto, A.; Marcomini, A. The KULTURisk Regional Risk Assessment methodology for water-related natural hazards—Part 1: Physical-environmental assessment. *Hydrol. Earth Syst. Sci.* **2014**, *18*, 5399–5414. [CrossRef]

9. Rizzi, J.; Torresan, S.; Zabeo, A.; Critto, A.; Tosoni, A.; Tomasin, A.; Marcomini, A. Assessing storm surge risk under future sea-level rise scenarios: A case study in the North Adriatic coast. *J. Coast. Conserv.* **2017**, *21*, 453–471. [CrossRef]

10. Gandini, A.; Garmendia, L.; San Mateos, R. Towards sustainable historic cities: Adaptation to climate change risks. *Entrep. Sustain. Issues* **2017**, *4*, 319–327. [CrossRef]

11. Grossi, C.M.; Brimblecombe, P.; Harris, I. Predicting long term freeze-thaw risks on Europe built heritage and archaeological sites in a changing climate. *Sci. Total Environ.* **2007**, *377*, 273–281. [CrossRef] [PubMed]

12. Huijbregts, Z.; Kramer, R.P.; Martens, M.H.J.; van Schijndel, A.W.M.; Schellen, H.L. A proposed method to assess the damage risk of future climate change to museum objects in historic buildings. *Build. Environ.* **2012**, *55*, 43–56. [CrossRef]

13. Rockman, M.; Morgan, M.; Ziaja, S.; Hambrecht, G.; Meadow, A. *Cultural Resources Climate Change Strategy*; Cultural Resources, Partnerships, and Science and Climate Change Response Program, National Park Service: Washington, DC, USA, 2016.

14. Bratasz, Ł.; Harris, I.; Lasyk, Ł.; Łukomski, M.; Kozłowski, R. Future climate-induced pressures on painted wood. *J. Cult. Herit.* **2012**, *13*, 365–370. [CrossRef]

15. Brimblecombe, P.; Lankester, P. Long-term changes in climate and insect damage in historic houses. *Stud. Conserv.* **2012**, *58*, 13–22. [CrossRef]

16. Lankester, P.; Brimblecombe, P. The impact of future climate on historic interiors. *Sci. Total Environ.* **2012**, *417–418*, 248–254. [CrossRef] [PubMed]

17. Klostermann, J.; van de Sandt, K.; Harley, M.; Hildén, M.; Leiter, T.; van Minnen, J.; Pieterse, N.; van Bree, L. Towards a framework to assess, compare and develop monitoring and evaluation of climate change adaptation in Europe. *Mitig. Adapt. Strateg. Glob. Chang.* **2018**, *23*, 187–209. [CrossRef] [PubMed]

18. National Park Service. Climate Change Response Strategy. 2010. Available online: https://www.nps.gov/subjects/climatechange/upload/NPS_CCRS-508compliant.pdf (accessed on 1 October 2018).

19. English Heritage. Climate Change and the Historic Environment. 2006. Available online: http://discovery.ucl.ac.uk/2082/1/2082.pdf (accessed on 1 October 2018).

20. Mahdjoubi, L.; Hawas, S.; Fitton, R.; Dewidar, K.; Nagy, G.; Marshall, A.; Alzaatreh, A.; Abdelhady, E. *A Guide for Monitoring the Effects of Climate Change. On Heritage Building Materials and Elements*; Report Prepared for the Funded Research Project: Heritage Building Information Modelling and Smart Heritage Buildings, Performance Measurements for Sustainability; British University in Egypt: El Sherouk City, Egypt, 2017.

21. National Research Council. *Monitoring Climate Change Impacts: Metrics at the Intersection of the Human and Earth Systems*; National Academies Press: Washington, DC, USA, 2010.

22. Leijonhufvud, G.; Henning, A. Rethinking indoor climate control in historic buildings: The importance of negotiated priorities and discursive hegemony at a Swedish museum. *Energy Res. Soc. Sci.* **2014**, *4*, 117–123. [CrossRef]

23. European Standard NS-EN 16096:2012. *Conservation of Cultural Property—Condition Survey and Report of Built Cultural Heritage*; British Standards Institution: London, UK, 2012.

24. Brimblecombe, P. Monitoring the Future. In *Climate Change and Cultural Heritage*; Lefèvre, R.A., Sabbioni, C., Eds.; Edipuglia: Bari, Italy, 2010.

geosciences

MDPI

Article

Indoor Multi-Risk Scenarios of Climate Change Effects on Building Materials in Scandinavian Countries

Arian Loli * and Chiara Bertolin

Department of Architecture and Technology, Norwegian University of Science and Technology,
7491 Trondheim, Norway; chiara.bertolin@ntnu.no
* Correspondence: arian.loli@ntnu.no; Tel.: +47-922-36-913

Received: 14 July 2018; Accepted: 12 September 2018; Published: 14 September 2018

Abstract: Within the built environment, historic buildings are among the most vulnerable structures to the climate change impact. In the Scandinavian countries, the risk from climatic changes is more pronounced and the right adaptation interventions should be chosen properly. This article, through a multidisciplinary approach, links the majority of climate-induced decay variables for different building materials with the buildings' capacity to change due to their protection status. The method tends to be general as it assesses the decay level for different building materials, sizes, and locations. The application of the method in 38 locations in the Scandinavian countries shows that the risk from climatic changes is imminent. In the far future (2071–2100), chemical and biological decays will slightly increase, especially in the southern part of the peninsula, while the mechanical decay of the building materials kept indoors will generally decrease. Furthermore, the merge of the decay results with the protection level of the building will serve as a good indicator to plan the right level and time of intervention for adapting to the future climatic changes.

Keywords: climate change scenarios; mechanical decay; biological decay; chemical decay; wood; masonry; Scandinavian countries; indoor climate

1. Introduction

The Scandinavian countries are predicted to be affected by climate change not only limited to the temperature increase [1]. In Norway, long-term climate projections up to the year 2100 have demonstrated that the country will face a significant increase of annual temperature, precipitations, floods, and mean sea level while the winter snow cover and the number of glaciers will be substantially reduced [2].

It is unavoidable that the climatic changes will affect humans and their living environment. The risk assessment of building materials and components serves as a basic step for defining the adaptation measures that need to be applied in buildings to adjust them to the "new" climate. In this context, several studies have been carried out to assess the decay level of the Scandinavian building stock induced by climate change regarding the type of constructive material: wood [3,4], masonry [5,6], and concrete [7].

Within the built environment, historic buildings are among the most vulnerable structures due to their relatively older age and un-renewable values that they represent. The report from UNESCO World Heritage Centre states that the impacts of climate change are affecting many and are likely to affect many more World Heritage sites, both natural and cultural in the years to come [8]. The topic has gained a lot of attention in the last decades with many studies focused on the intersection between the climate change and cultural heritage management [9]. The objective of these research studies is to assess the potential impact of the climate change in heritage sites and propose strategies to face the future risks [10–12]. In addition, many international projects at European level have been running or are ongoing with the primary goal to alleviate the negative effects of the decay induced by climatic changes [13,14].

Due to its severity, the impact of the climate change has obtained high attention in the Scandinavian cultural heritage sector where intergovernmental meetings with a focus on conservation, planning, and management of the cultural environment have been held [15]. Their scope is to assist the cultural heritage managers in adapting to climate change and to strengthen the collaboration between the Scandinavian countries [16]. To this aim, a step-by-step methodology is proposed in this article for helping the heritage owners and managers to evaluate possible climate-induced risk on building materials and take precautions against it [17].

The achievements of the materials science researchers and cultural heritage specialists regarding the effects of climate change are important, but they should be merged to find suitable adaptation interventions that satisfy the demands of both communities. The scope of our article is to link the majority of climate-induced decays that can affect historic buildings with the level of legislative adaptation intervention (small, medium, large) allowed to them in one multidisciplinary method. The use of available data from the European project Climate for Culture (CfC: 2009–2014) [18] is enhanced in the proposed method. The data are used to estimate the total level of decay in a range of 16 buildings, with different sizes and construction materials, with the purpose to quantify the comprehensive effects of the expected climate change in the far future (2071–2100). The simulations provide information to cultural heritage managers and help the stakeholders to understand the type of structure that resists better, in natural conditions, to climate change impact and the geographical locations that are more exposed to risk in the Scandinavian region. This type of information is extremely valuable because, after the merge with the protection level of the building, it serves as a good indicator to plan adequate adaptation interventions and implement them with the necessary level of urgency.

2. Materials and Methods

2.1. Climate for Culture Project

The European project Climate for Culture, which investigated the potential impact of climate change on Europe's cultural heritage assets, particularly on historic buildings, provided high-resolution risk maps that identify the most urgent outdoor risks for European regions until 2100 but also risks for indoor collections [13]. These maps are the output of climate change scenarios coupled with building simulations at the European scale and serve as a powerful tool for preventive conservation and decision-makers that deal with cultural heritage [19]. The maps, through the colour codes, show the level of risk, both for outdoor and indoor environments, for 16 building types and 19 environmental variables. The results of the project can be used to understand how the climatic changes affect the buildings in natural conditions (without the use of indoor heating, ventilation, and air conditioning (HVAC) systems) in relation to their geographical location, building size, window size, and constituting material.

2.1.1. Description of Buildings and Climatic Data

In the CfC project, the general assessment and map creation process has been carried out using the following specifications:

- Emission scenario

The impact of climate change on historic buildings was evaluated using the high-resolution regional climate model REMO [20] which provides climate change projections for entire Europe at 12.5 km spatial resolution. Two emission scenarios were applied in the project. The first is the mid-line A1B scenario [21], which considers a CO_2 emission increase until 2050 and a decrease afterwards. The second is the more recent Representative Concentration Pathway 4.5 Emission

Scenario (RCP4.5) of the Intergovernmental Panel on Climate Change (IPCC) assessment report 5 (AR5) [22]. This scenario is based on long-term, global emissions of greenhouse gases, short-lived species, and land-use-land-cover, which stabilizes radiative forcing at 4.5 watts per metre squared (approximately 650 ppm CO_2 equivalent) in the year 2100 without ever exceeding that value.

- Locations

Climate data assessment and simulations were calculated for a regular grid that covers entire Europe including the Mediterranean region.

- Time windows

The climate data were produced for all the climate-induced variables from hourly data elaboration over two 30-year time windows: 2021–2050 (Near Future) and 2071–2100 (Far Future), maintaining the period 1961–1990 (Recent Past) as a reference period (Table 1).

Table 1. The time windows used for simulations in the Climate for Culture project.

Recent Past (RP)	Near Future (NF)	Far Future (FF)
1961–1990	2021–2050	2071–2100

- Future indoor climates and risk assessment

The outdoor climate influences the cultural heritage structures, both in terms of outdoor and the indoor environment [13]. The future climate predictions explained above were used to create the risk maps for the outdoor environmental variables, which provide important information for decision makers to plan outdoor adaptation measures. These climate change predictions linked with building simulations allow the estimation of indoor climate variables (temperature T, relative humidity RH) and indoor damage variables for mechanical, chemical and biological decay using an automated method [23]. The risk induced indoor by climate change is assessed by the combination of indoor climate data with the damage functions of the variables [24].

- Buildings

Indoor climates of historic buildings were modelled and simulated following two different approaches. The first consisted of the development of a full-scale multizone dynamic hygrothermal whole building simulation while the second used a simplified hygrothermal building model. The first model gives more detailed results about the temperature and relative humidity inside the building, but it has a high development cost and takes long simulation time. The simplified model provides reliable results within a short simulation time and for this reason, it was applied in the CfC project to predict indoor temperature and relative humidity. It has the restriction to be effective to buildings without active HVAC systems and to request all the necessary measured values for the parametrisation of the model [13]. Through this model, it was possible to perform simulations on 16 generic sacred buildings, virtually located in all the grid cells, for producing indoor climate data and risk maps. The general layout of buildings is composed of a rectangular floor plan, a gable roof, and long walls in the North-South direction with windows only on the long walls. Each of the buildings is unconditioned and their matrix is a combination of their volume (small/large), window area (small/large), structure (heavyweight/lightweight), and moisture buffering capacity (MBP) (low/high) as given in Table 2.

Table 2. Generic sacred building matrix.

		Heavyweight		Lightweight	
		Low MBP	**High MBP**	**Low MBP**	**High MBP**
Small Building	**Small Window**	Building 1	Building 2	Building 3	Building 4
	Large Window	Building 5	Building 6	Building 7	Building 8
Large Building	**Small Window**	Building 9	Building 10	Building 11	Building 12
	Large Window	Building 13	Building 14	Building 15	Building 16

2.1.2. Indoor Deterioration Variables

The variables, according to the CfC results, for each mechanism of indoor deterioration (mechanical, chemical, and biological) and assigned to different building materials (wood, masonry, and concrete), are reported in Tables 3 and 4 [25].

Table 3. The main variables to estimate indoor deterioration in wooden buildings.

Mechanical Damages	Chemical Degradation	Biological Deterioration
Panel (base material)	Lifetime multiplier	Mould
Panel (pictorial layer)		Insects
Jointed element		
Cylindrical element		

Table 4. The main variables to estimate indoor deterioration in masonry and concrete buildings.

Mechanical Damages	Chemical Degradation	Biological Deterioration
Salt crystallisation cycles	Lifetime multiplier	Mould
Thenardite-Mirabilite cycles [1]		
Freeze-thaw cycles		
Frosting time		

[1] Only for concrete structures.

A short explanation for the indoor decay-linked mechanisms, according to the CfC deliverable D4.2 [24], is given as follows:

- Mechanical risk for wooden elements: panels, jointed elements, cylindrical elements

The *RH* of the air affects the moisture content (MC) in a wooden element. As the moisture content changes, so do the dimensions of the element, which set up internal stresses that lead to deformations. At low stresses, the wood behaves elastically, with reversible deformations while above a certain threshold of strain (the yield point), the deformation becomes plastic, the change is not reversible anymore and the material fails. The damage functions used in CfC for this type of elements are based on Marco Martens' interpretation [26] of studies by Mecklenburg, Bratasz, and Jakiela [27–29]. Different response times are used in the algorithm to smooth out the *RH* fluctuations in order to represent better the moisture changes experienced in the substrate of different building elements in wood. The strains induced by the expected changes are calculated and a final assessment is made to evaluate if the resultant strain falls in the area of elastic (green code), plastic (orange code), or failure (red code) response.

- Mechanical risk for masonry and concrete

Salt-crystallization cycles. Damage from salt crystallization occurs at the interface between air and the object, or beneath the surface of the object. The surface gets covered by a mass of small crystals that destroy the visual integrity or disfigure the natural appearance of masonry or concrete. When this occurs below the surface, the visible result is surface disruption and loss of material. The damage function for stone weathering is studied from Grossi et al. [30] and predictions in the context of climate change are discussed in the atlas of Noah's Ark project [31] and reported by Lankester and Brimblecombe [32]. The damage function used in CfC for calculating the number of cycles counts the transition that occurs in a range around 75% *RH* (independently from the temperature) as this is the threshold of deliquescence of the sodium chloride.

Thenardite-Mirabilite cycles. Similarly, the porous stone might be destroyed due to the pressure exerted during the transition from the thenardite (Na_2SO_4) to the mirabilite ($Na_2SO_4 \times 10H_2O$) that occurs with the inclusion of 10 molecules of water in the hydrated crystal. Mirabilite exerts a very high crystallization pressure on the porous wall causing the damage of stone. A pressure of about 10 MPa occurs when the *RH* increases across value described by a critical $RH = 0.88 \times T + 59.1$. Repeated cycles may accumulate stress and in the long-term, they may cause severe decay. The damage function used in CfC counts the transition that occurs in the thenardite-mirabilite system and estimates a green code for up to 60 cycles; orange code for 60–120 cycles and red code over 120 cycles.

Freeze-thaw cycles. When water goes from liquid to solid phase within a porous masonry element or in a structural crack, it increases in volume, which can cause damaging stress. If this stress is repeated in a cyclic way, the brick or stone becomes weaker, and eventually delaminates and spalls. The theoretical background of freeze-thaw cycles is discussed by Camuffo [33] and in the atlas of Noah's Ark project [31]. The damage function counts the number of cycles between $T < -3\,°C$ (freeze) and $T > +1\,°C$ (thaw) that occur in one year. The results of CfC maps indicate a green code for up to 30 freeze-thaw cycles, orange code for cycles between 30 and 60 and red code for more than 60 cycles during the year.

Frosting time is considered the total amount of time (in hours) during the year when the air temperature (outdoor or indoor) is below zero degrees Celsius. The effect of frosting time over cultural heritage materials has been studied by Camuffo [33]. Separately, this variable is not helpful to predict material damage but it may serve as an indicator for further investigations. Frosting time can be a useful parameter in sub-zero temperature zones (many zones in the Scandinavian countries) where it determines the penetration risk of the ice front through the building wall. The level of risk according to CfC maps is estimated green for up to 2400 h/year, orange for frosting time between 2400 and 4800 h/year, and red for more than this value.

- Chemical risk

Lifetime Multiplier (LM) is the ratio between the predicted lifetime of the material subjected to the environmental conditions and the predicted lifetime at standard conditions of 20 °C and 50%RH. When *LM* > 1, the material will last longer than the standard conditions (green code) while for *LM* < 1, the rate of deterioration is greater and the lifetime shorter. The level of *LM* < 0.5 (half of lifetime), is defined as the threshold of high risk and is illustrated in red.

The calculation of the *LM* for different types of materials is done using the Equation (1) derived by Michalski [34]:

$$LM = \left(\frac{50\%}{RH}\right)^{1.3} \times e^{\frac{E_a}{R}\left(\frac{1}{T} - \frac{1}{293}\right)} \tag{1}$$

where *RH* is the relative humidity [%], *T* is the absolute temperature [K], E_a is the activation energy [J mol^{-1}], and *R* is the constant of gas (8.314 [J mol^{-1} K^{-1}]).

In the calculations, the value of activation energy (the least possible amount of energy which is required to start a reaction) is considered 59.24 kJ mol^{-1} for wood and 42.5 kJ mol^{-1} for masonry and concrete. The values are taken as average because the activation energy can vary for a different range of materials. The equation does not consider the effects of very low or very high *RH* but it can be a good indicator of the decay rate if the *LM* will increase or decrease in the future.

- Biological risk

Mould growth is an extensive problem that implicates the human health and the integrity of the material. The effects on heritage items can vary from light powdery dust to severe stains, which weaken and disintegrate the substrate material. It is assumed that at temperatures above zero degrees Celsius and relative humidity above 70% the mould spores can germinate. The rate of growth depends on the climatic conditions, type of material but also the accumulation of dirt and dust in case of inorganic materials. The CfC maps have been developed using the Sedlbauer isopleths system [35] and they consider a growth rate of less than 50 mm/year as safe (green), a growth rate between 50 and 200 mm/year as possible damage (orange), and an annual growth rate greater than 200 mm as damage (red).

Insects can be another cause of damage to heritage items. The damage can be caused by certain moths and beetles and some forms of insects such as silverfish and booklice. The risk of damage from insects depends on relative humidity for some species and on temperature for most insect types. The key factors in assessing risk are climatic conditions, type of insect, and the vulnerability of the organic material such as wood. The insects' activity is present in temperatures of 5–30 °C but below 15 °C, their damage is limited [36]. The results for the CfC project have been achieved by calculating the annual degree-days over 15 °C (($days \times (T - 15)$)) with $RH > 75\%$ and $T < 30$ °C for humidity dependent insects and $T < 30$ °C for temperature dependent ones.

2.2. Risk Assessment

The tool presented here tends to be general as it assesses the total decay level for the building materials. For this reason, the results of all decay-linked variables explained in the Section 2.1.2 and simulated in the CfC project, will be used as input to assess the overall deterioration of different building materials. The level of decay for each variable is divided into 6 category levels: very low, low, medium, medium-high, high, and very high. The threshold values for each decay level are established from the description of the variables in the CfC deliverable D4.2 [24] and the colour code of the risk maps from the project output which considers the likelihood and the impact of the decay. The boundary value for each level is shown in Table 5.

Table 5. The table of risk assessment for the main deterioration variables.

Variable Name	Unit	Very Low	Low	Medium	Medium-High	High	Very High
Panel—base material	[-]	0.333	0.667	1	1.333	1.667	2
Panel—pictorial layer	[-]	0.333	0.667	1	1.333	1.667	2
Jointed element	[-]	0.333	0.667	1	1.333	1.667	2
Cylindrical element	[-]	0.333	0.667	1	1.333	1.667	2
Salt crystallisation cycles	[no/year]	30	60	90	120	150	180
Thenardite-Mirabilite cycles	[no/year]	30	60	90	120	150	180
Freeze-thaw cycles	[no/year]	15	30	45	60	75	90
Frosting time	[h/year]	1200	2400	3600	4800	6000	7200
Lifetime multiplier—Wood	[-]	1.5	1.25	1	0.75	0.5	0.25
Lifetime multiplier—Masonry	[-]	1.5	1.25	1	0.75	0.5	0.25
Lifetime multiplier—Concrete	[-]	1.5	1.25	1	0.75	0.5	0.25
Mould growth	[mm/year]	25	50	125	200	400	600
Insects—humidity dependent	[DD/year]	500	1000	1500	2000	2500	3000
Insects—temp. dependent	[DD/year]	500	1000	1500	2000	2500	3000

Firstly, the level of risk is weighed for each decay-linked variable using the thresholds given in Table 5. In the second step, depending on the constituting building material, the risk is evaluated for each mechanism of deterioration: mechanical, chemical, and biological. When more than one decay-linked variable is needed to evaluate the level of deterioration of a specific mechanism (e.g., mechanical decay), the highest risk level among the variables determines the risk level of the entire mechanism. This assumption has been made by assigning the same importance to each-decay linked variable due to their likelihood and associated impact. The third and last step is the assessment of the total level of decay of the building, based on the rating of the three deterioration mechanisms computed in the second step. The same assumption as in the previous step is used, i.e., the mechanism with the highest level of risk determines the total level of decay of the building.

2.3. Historic Significance Assessment

In parallel with the decay assessment, the other stage that deals with the assessment of the historic values of the buildings, should be performed. While the first stage covers the technical and physical characteristics of the building, the significance assessment highlights the social, artistic, and cultural aspects of it. The assessment of the character-defining elements (CDEs) is very important prior to take adaptation actions in historic buildings because it safeguards the values that need to be preserved and avoids incorrect or irreversible interventions [37].

On this background, a tool named DIVE (Describe, Interpret, Valuate, and Enable) has been developed for assessing the historic significance of buildings and suggesting the potential field for actions/interventions [38]. The method is a result of two international projects "Sustainable Historic Towns: Urban Heritage as an Asset of Development" (SuHiTo: 2003–2005) [39] and "Communicating Heritage in Urban Development Processes" (Co-Herit: 2007–2008) [40] with partners from Finland, Lithuania, Norway and Sweden and it emphasizes the importance of collaboration between cultural heritage professionals and decision-makers.

The name of the method is an acronym of the four main stages of it that are connected like links in a chain (Figure 1). DIVE is an interdisciplinary and participatory methodology that involves different target groups from both the public and private sector. The tool has been applied to different cultural environments in North Europe like in the towns of Jakobstad in Finland, Naujoji Vilnia in Lithuania, Odda and Tromsø in Norway, Arboga in Sweden, etc. [40].

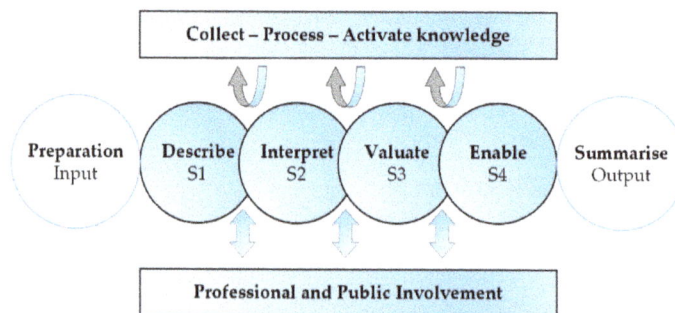

Figure 1. Structure of the DIVE approach.

The output of the method enhances the simultaneous importance of preserving social, cultural, and physical features of the buildings in the future development of historic urban districts by stating the attributes that carry a primary role and those that are of secondary importance. The recommendations are given for every attribute (shape, windows, ceiling, stairs, walls, etc.) that are grouped into four main categories: exterior, interior, structure, and use of the building. The analysis tends to categorise the buildings according to the values that they represent as well as the scale of interventions (capacity to change) allowed on them [38].

According to DIVE output, the capacity of a building to change can be of a small, medium or large scale (e.g., preservation of the window frame, replacement by keeping the same format and proportions or replacement with a new window). Meanwhile, during the application of the method in case studies, the applied grading system results with six intervention levels by adding also intermediate levels: none to small, small, small to medium, medium, medium to large, and large corresponding to the levels of protection: very high, high, medium-high, medium, low, and very low [41].

3. Results

3.1. Matrix of Selection of Adaptation Intervention

Finding the best adaption intervention scenario in historic buildings is a complex process because it has to boost the historic value of the building, to decrease the damage of the decay processes, and, at the same time, to satisfy the increasing demands regarding the minimisation of carbon footprint and energy use. For this reason, the intervention should take into consideration three important parameters: level of protection that safeguards the significance of the historic building, level of decay in the building and the environmental impact of the intervention by minimizing the use of new materials and energy. The environmental impact has a substantial contribution towards the minimization of the climate change impact and should be considered throughout the selection of the adaptation intervention [37].

A possible adaptation scenario in an historic building should be able to respond properly to the expected level of climate-induced decay. In addition, the level of intervention should decrease the expected damage in a conservative approach. This can be achieved by linking the results of the risk assessment (Section 2.2) with the historic value assessment (Section 2.3). The proposed matrix that connects the levels of decay with the levels of legislative protection in historic buildings is given in Figure 2.

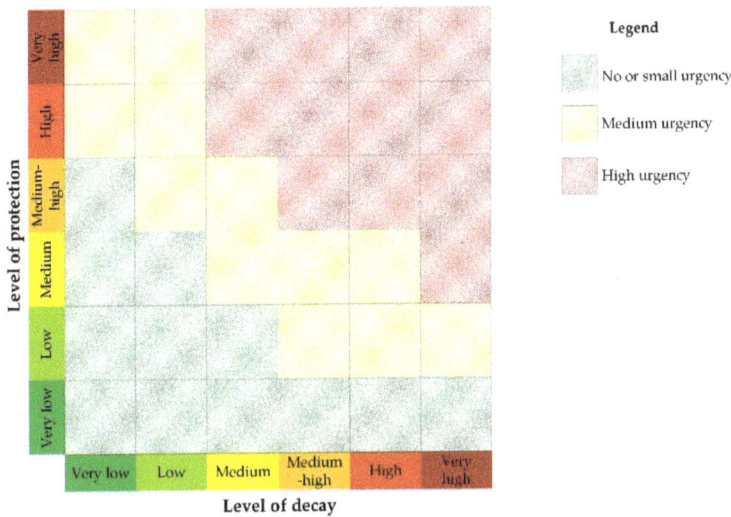

Figure 2. The link between the levels of decay and the levels of protection in historic buildings. The colour code highlights the urgency need for adaptation interventions.

The matrix is an useful tool for the decision-makers because it represents the limits of effectiveness of an adaptation intervention by considering both the level of deterioration and the scale of intervention allowed to the structure. Figure 2, through the colour codes, indicates the urgency needs before planning and implementing adaptation interventions. The riskiest situations (red nuances in the right-upper side of the matrix) can occur when a very valuable historic building is subject to natural hazards or catastrophic events such as earthquake, fire, floods, wars, etc. or in heavy conditions due to continuous disuse and lack of maintenance. In such cases, the legislative requirements of small interventions (e.g., ordinary maintenance and cleaning) cannot solve the strong symptoms of decay and therefore, higher level of interventions is required with urgency. The intervention target should be primarily directed towards the stability of the structural elements in order to avoid the collapse of the whole structure and the loss of the cultural heritage.

When the deterioration level is high or very high, the judgement can confirm (green nuance) or exceed (yellow and red nuances) the scale of the allowed interventions, depending on the protection category of the building [42,43]. However, examples of wrong, heavy or useless invasive interventions on architectural heritage sites exist after disastrous events [44] or as a result of wrong decision-making processes (e.g., the refurbishment case of the Matrera castle in Spain where smooth concrete walls were added to the original stone structure). Safety interventions, necessary to avoid collapses during the aftershocks or long disuse of the building, can hide or reduce the value of the original historic building when no compatible or durable materials are used. After such interventions, the return of the structure to the original form can be more difficult and expensive.

When the level of decay does not affect the load-bearing capacity of the structure but comes as a result of natural weathering (up to medium-high decay), the selected intervention should maintain the historic attributes of the building, through the applications of both preventive conservation measures and non-destructive interventions [45,46].

In the left-lower side of the matrix, the green area reports the "ideal" situation, i.e. when the building itself has not many CDEs at risk and/or when the decay level is not high to be kept under control using conservative interventions. However, bad conservative practices can fall even in the green area of Figure 2. These overdoing practices, common when adapting a historic building to modern use (e.g., change of use or capacity) or to new comfort requirements, do not always fit with the original design of the building and have the additional risk to use unnecessary economic and environmental resources.

3.2. Influence of Climatic Changes to Future Interventions in the Scandinavian Countries

The climate change effect will affect the deterioration level in historic buildings depending on the geographic location and type of constructive material [31]. For this reason, cultural heritage managers have to plan and implement adaptation actions that can work effectively for the years to come [47]. An effective adaption intervention has to consider not only the actual situation of the building but also the effects of the climate change over an extended period. In the following example, the level of decay is estimated over two time windows: the Recent Past (RP) and the Far Future (FF) to evaluate the expected effect of climatic changes over building materials. Thirty-eight locations in the Scandinavian countries are extracted from the general European and Mediterranean grid of the CfC project. The coordinates and the labels of each location are provided in Table 6.

For each location, data from the 16 generic sacred unconditioned buildings are taken from the CfC project simulations in term of indoor conditions. In the project, the choice of working with scenarios in unconditioned buildings (without indoor HVAC systems) was made because the climate change effects can be more clearly identified indoors and due to the limitations of the simplified simulation method. The values of the variables are taken from the RCP4.5 emission scenario, because it is the most recent one.

3.2.1. Decay-Linked Variables

Charts that visualize the connection between the decay-linked variables and the geographic location of the buildings are created for the RP and FF time windows and collected in Supplementary Materials. In the charts, only the set of dots has a numerical meaning; however, the dots of the same building are connected with lines and colour codes to facilitate the reading and allow distinguishing the values among buildings sizes and materials. The charts in Supplementary Materials are presented for each climate-induced variable in relation to the time window, geographical location, material, and size of the building.

Table 6. The map and the coordinates of the 38 locations extracted from the Climate for Culture project (image generated from [48]).

Map	ID	Lat.	Long.	Country
	1	69.2898	17.5711	Norway
	2	69.2659	21.1079	Norway
	3	69.1791	24.6277	Norway
	4	67.9698	17.5926	Sweden
	5	67.9471	20.9522	Sweden
	6	67.8647	24.2975	Finland
	7	66.6144	14.4121	Norway
	8	66.6498	17.6118	Sweden
	9	66.6282	20.8133	Sweden
	10	66.5496	24.0026	Finland
	11	65.2960	14.5712	Sweden
	12	65.3298	17.6291	Sweden
	13	65.3092	20.6884	Sweden
	14	63.8928	11.7972	Norway
	15	63.9774	14.7152	Sweden
	16	64.0099	17.6447	Sweden
	17	63.9901	20.5754	Sweden
	18	62.2671	6.5031	Norway
	19	62.4467	9.2598	Norway
	20	62.5775	12.0436	Sweden
	21	62.6587	14.8462	Sweden
	22	62.6899	17.6589	Sweden
	23	60.9630	6.9321	Norway
	24	61.1359	9.5886	Norway
	25	61.2618	12.2690	Norway
	26	61.3399	14.9660	Sweden
	27	59.6575	7.3273	Norway
	28	59.8244	9.8912	Norway
	29	59.9457	12.4764	Sweden
	30	60.0210	15.0762	Sweden
	31	60.0499	17.6838	Sweden
	32	58.3510	7.6928	Norway
	33	58.7020	15.1779	Sweden
	34	57.1994	10.4305	Denmark
	35	57.3127	12.8455	Sweden
	36	57.3829	15.2722	Sweden
	37	55.7352	8.3487	Denmark
	38	55.9958	13.0108	Sweden

3.2.2. Level of Decay

Wood is the dominant structural material of the constructions in the Scandinavian countries. The decay assessment for wooden buildings is computed using the CfC data related to lightweight buildings for both small (3, 4, 7, 8) and large (11, 12, 15, 16) building sizes (see Table 2). For structures in masonry or concrete, the decay level is assessed using the CfC data related to heavyweight buildings regarding the two size groups: small (1, 2, 5, 6) and large (9, 10, 13, 14). The decay assessment is performed for each group of four buildings using the methodology described in Section 2.2, for both the RP and FF. Given a specific location, the level of a decay-linked variable is evaluated considering the highest value of the variable within the group of four buildings that represents a specific building material and size.

The results of the risk assessment for each mechanism of deterioration (mechanical, chemical and biological) and for the total level of risk regarding small/large and light/heavyweight buildings in the two time windows are reported in Figures 3–6, using the risk assessment colour code (Table 5).

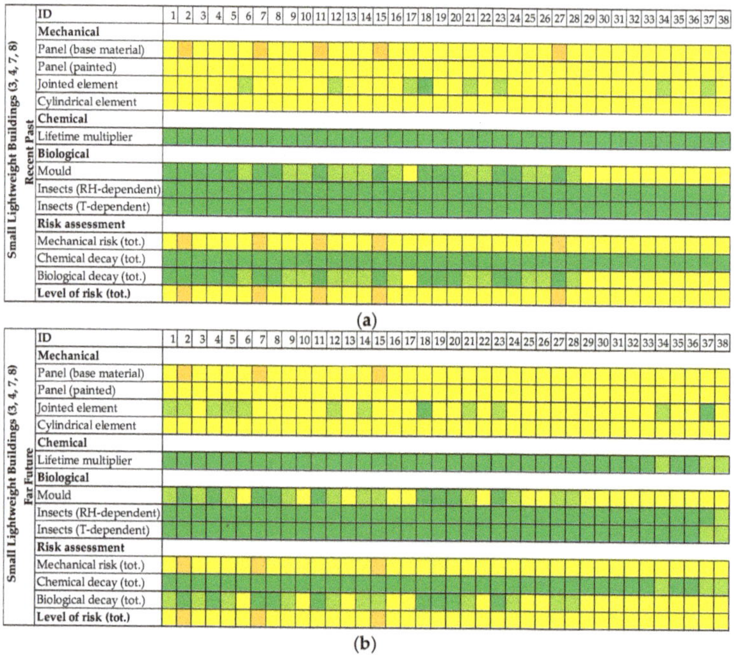

Figure 3. Risk assessment matrix of the deterioration of small lightweight buildings in: (**a**) Recent Past (1961–1990); (**b**) Far Future (2071–2100).

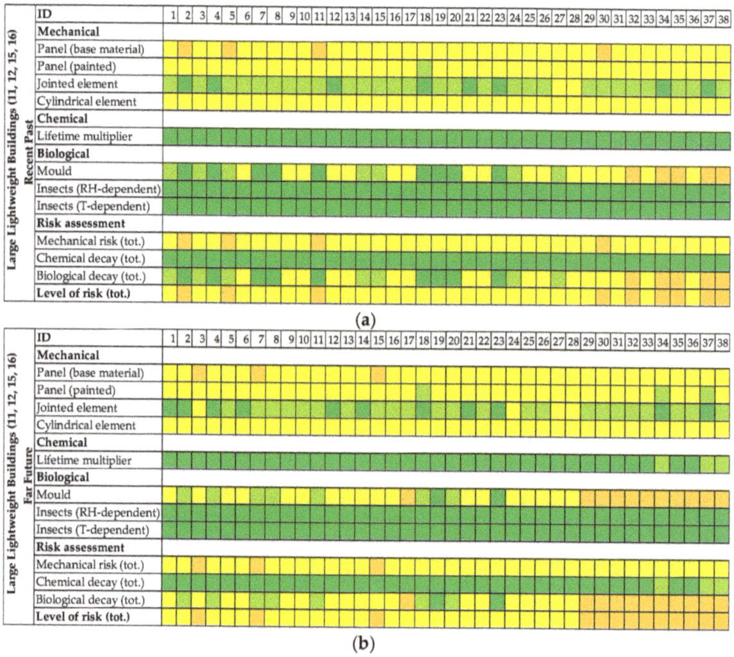

Figure 4. Risk assessment matrix of the deterioration of large lightweight buildings in: (**a**) Recent Past (1961–1990); (**b**) Far Future (2071–2100).

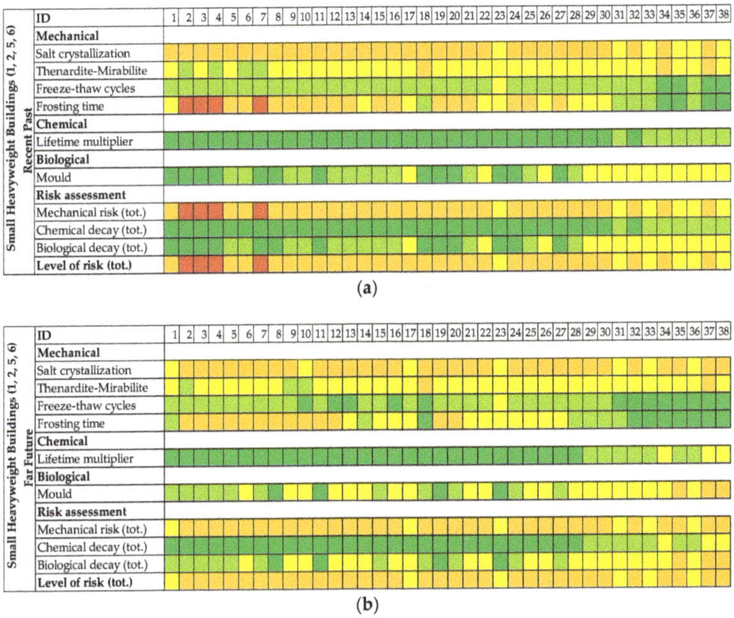

(a)

(b)

Figure 5. Risk assessment matrix of the deterioration of small heavyweight buildings in: (**a**) Recent Past (1961–1990); (**b**) Far Future (2071–2100).

(a)

(b)

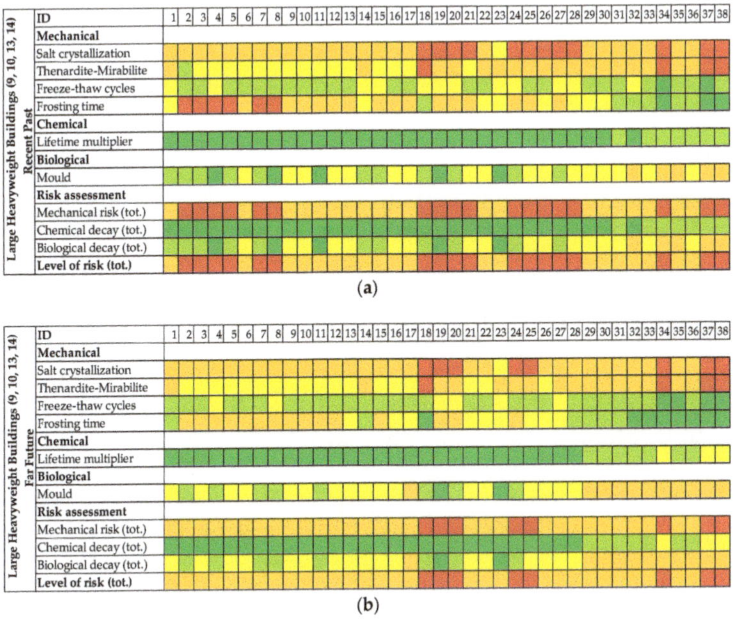

Figure 6. Risk assessment matrix of the deterioration of large heavyweight buildings in: (**a**) Recent Past (1961–1990); (**b**) Far Future (2071–2100).

From an overview analysis of the graphs in Supplementary Materials and the risk matrices, which summarize the single types of risk and the total level of risk, it is noticed the following:

- In Scandinavian countries, the mechanical deterioration indoor in all types of buildings has a general decrease in the Far Future, although the decay remains in the ranges of medium to high risk.
- The chemical and biological risks increase. The former, exemplified by the lifetime multiplier indicator, remains in the range of low decay, except for the last locations in the map, corresponding to the south of Scandinavian Peninsula and Denmark (ID: 29–38) where the risk increases to a low or medium level. In the other ID points, the increase is still distinguishable, but it remains within the same level of risk for the buildings.
- The biological risk increases the number of locations in which the decay will fall in low, medium, and medium-high, especially in western Sweden, south of Scandinavian Peninsula and Denmark.
- Regarding the risk level over the Recent Past (1961–1990), high level of risk (red colour) in the indoor environment is noted only for heavyweight buildings, which resemble the masonry or concrete constructions. This level is caused by the mechanical damage in the building materials and has a throughout geographical distribution: in northern Scandinavian Peninsula (ID: 2–5, 7–8) due to the frosting time, while in central and southern parts of the peninsula (ID: 18–21; 24–28; 34; 37–38) due to the salt crystallization by sodium chloride and the transition from thenardite to mirabilite. In the Far Future (2071–2100), the risk tends to decrease because of the climatic changes, e.g., the risk due to frosting time in the northern peninsula will have a transition from high to medium level (ID: 2–5, 7–8).
- Specific effects of the climate change are also noted when the sizes of the buildings are compared. In all the three deterioration mechanisms, the level of risk in large buildings (ground floor area larger than 320 m^2) results higher in comparison with small buildings, regardless the time window, constituting material, and the geographical location. In the Far Future, the decay risk of large lightweight buildings in the southern part of the peninsula (ID: 29, 32, 34–35, 37–38) will be medium-high due to climate change while small buildings in the same locations will face medium risk. Regarding heavyweight buildings in the Far Future, large ones will be disposed to high risk in central and southern areas (ID: 18–20, 24–25, 34, 37–38) while small buildings in these areas will remain at medium-high risk level.

3.2.3. Level of Intervention

The overall scenarios of the climate change effects on building materials, reported in each multi-risk table, can be used from the stakeholders to choose the urgency of the adaptation interventions that need to be implemented on historic buildings. This is achieved by linking the level of total decay of the buildings with the level of protection and adaptation interventions permitted by law. By applying the matrix in Section 3.1, the stakeholders can compare the actual urgency level (RP) on specific building materials, sizes, and locations in Scandinavian countries with those expected over the Far Future.

The locations in the Scandinavian countries, where interventions are required to minimize the risk of losing CDEs in historic heavyweight buildings, are inserted in the matrices in the Figures 7 and 8. The level of protection of the buildings according to the legislation is considered medium and medium-high in all the locations, which resembles small to medium and medium capacity to change.

Figure 7 shows the urgency levels of small heavyweight buildings to adapt measures that minimize the decay over the RP (Figure 7a) and the FF (Figure 7b) in relation to their ID locations. Over the RP, decay conditions in most of the locations (except ID: 23, 31, 33, 35, 36, 38) require adaptation measures to be implemented with high urgency when the level of building protection is medium-high. Over the FF, the buildings will experience a total decay reduction, with a consequent lower need for urgent adaptation interventions. At the opposite, small heavyweight buildings located near Göteborg and Malmö (ID: 35, 38, underlined) will shift from medium to medium-high risk of losing CDEs, requiring higher priority in adaptation and a higher level of intervention than those needed actually to counteract the decay.

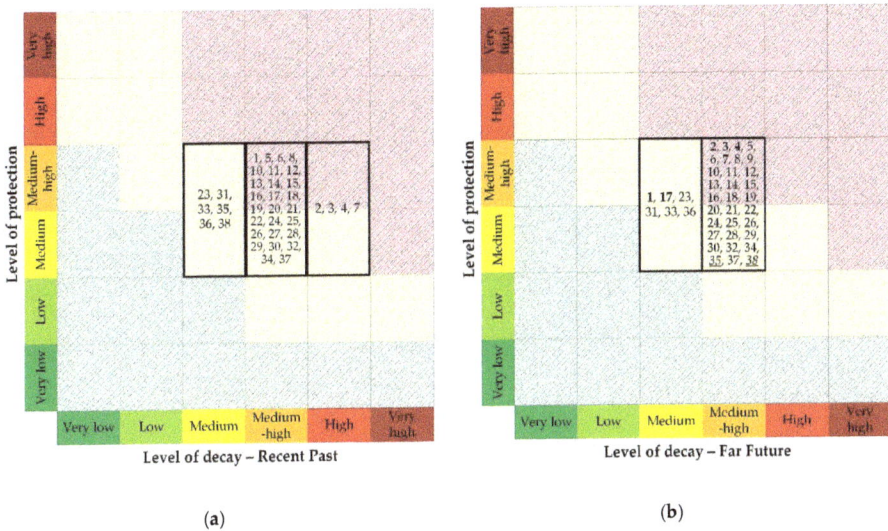

(a)　　　　　　　　　　　　　　　　　　(b)

Figure 7. The urgency of interventions in small heavyweight buildings for each ID location in: (a) Recent Past (1961–1990); (b) Far Future (2071–2100). (Underlined locations in FF: higher decay induced by climate change. Bold locations in FF: lower decay induced by climate change).

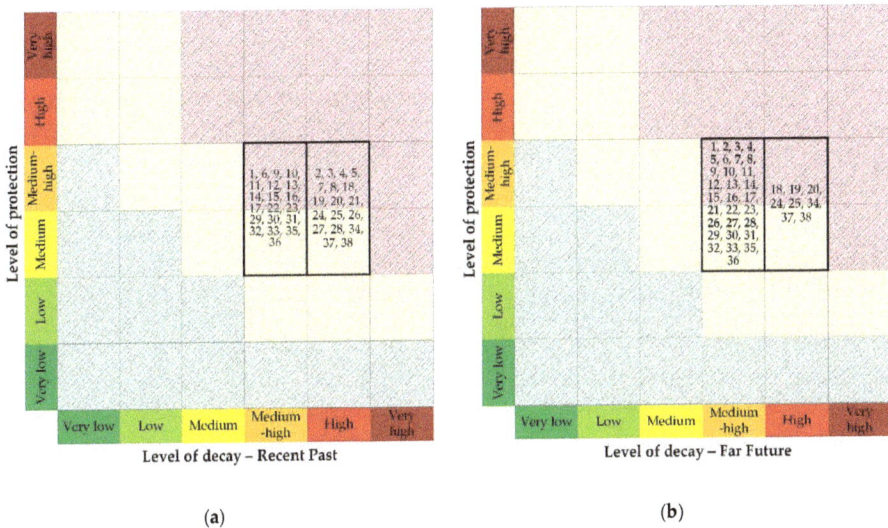

(a)　　　　　　　　　　　　　　　　　　(b)

Figure 8. Urgency of interventions in large heavyweight buildings for each ID location in: (a) Recent Past (1961–1990); (b) Far Future (2071–2100). (Bold IDs in FF: lower decay induced by climate change).

Figure 8 demonstrates the urgency of adaptation interventions that need to be applied on large heavyweight buildings in the 38 locations, for both the RP and FF climate-induced decay scenarios.

In Figure 8b, the adaptation interventions over the Far Future will have the same class of urgency as during the Recent Past but the interventions will be proposed for a lower level of decay in ID locations: 2–5, 7, 8, 21, 26–28 (in bold). From a comparison between Figures 7b and 8b, the decay level in the Far Future will remain extensively medium-high for both building sizes but in some locations it will be high for large buildings (ID: 18–20, 24, 25, 34, 37, 38) and medium for small buildings (ID locations: 1, 7, 23, 31, 33, 36).

4. Discussion

The proposed method at Section 2.2 has been applied to assess the total risk of climate-induced decay on building materials preserved in an indoor unconditioned environment of different dimensions. The thresholds, used in the quantification of the decay level, have been defined using the CfC risk assessment method that evaluates the impact and the likelihood of different types of decays through the use of damage functions. The threshold values are average and they can vary for a different range of materials due to their physical and mechanical characteristics or aggressiveness of the environment.

The same approach can also be feasible for assessing the risk of decay outdoors using the outdoor CfC maps and the variables that better estimate damage mechanisms induced by climate and weather conditions.

The main objective of the matrix proposed in this article is to find suitable adaptation interventions that fulfil both the physical state of the original material (to reach a minimum level of decay) and its historic value (to minimize the risk of losing CDEs). Using the proposed matrix, three types of intervention needs can be identified as follows:

1. No or small urgency of adaptation interventions rather than those allowed by the legislation (green colour in the matrix). This level is expected for existing buildings that are neither listed nor protected, as they have no specific need to guarantee the conservation of CDEs.
2. Medium urgency of adaptation interventions (yellow colour in the matrix) is expected for historic buildings that are listed. Within this category of buildings, the ongoing climate change effect will require, in the next decades, implementation of different levels of intervention than those admitted by the legislation, for responding effectively to the expected decay.
3. High urgency of adaptation interventions (red colour in the matrix) is expected for fully protected historic building, monuments and UNESCO sites. Within this category, new adaptation interventions have to be planned and implemented to respond both to the preservation of their valuable CDEs and to intervene with urgency in post-disaster situations.

Within the same matrix cell, more than one adaptation action can be recommended. In this case, the life cycle assessment (LCA) method can be applied as an effective decision-making tool to choose eventually the scenario with the lowest environmental impact. This decision leads to the choice of the greenest intervention, thus avoiding contributing to further the climate change. The environmental assessment can be a consequent component that completes the intervention selection process on historic buildings. Considering carbon footprint of the intervention reduces the impact of the climate change and makes the entire process three-dimensional where each component (level of decay, level of protection and level of emissions) is independent of each other but a correct combination of them achieves satisfactory results to answer the needs of cultural heritage preservation in the time of climate change.

5. Conclusions

Due to climate change impact, the cultural heritage management sector will face new challenges in the future (e.g., more info on identification, documentation and mapping of heritage sites with increased vulnerability to climate change will be needed). The main aim of the presented article is to enhance the use of already existing Climate for Culture results in order to create a tool (matrix) that provides information to cultural heritage managers regarding the urgency of intervention and the effectiveness of measures supported by legislation in reducing the level of decay.

The merge of the expected decay results with the level of protection of the building serves as a good indicator to enhance the reaction capacity and to plan the right level and time of intervention for adapting to the future climatic changes. By directing the adaptation intervention process in a methodologic approach, what today is a subjective choice, taken on a case-by-case basis, would become a more scientific and technical assessment.

The application of the method in 38 locations in the Scandinavian countries shows that the increase of temperature and relative humidity throughout the region will increase the conditions for biological growth of mould and insects as already confirmed from other researches in the field [16]. This risk is imminent in the region where about 90% of the structures and the majority of historic buildings are built from wood, especially in the southern areas where the climatic conditions are more favourable for growth. The climatic changes will affect also the structures that have iron elements due to the increase of risk for corrosion. While the biological and chemical deteriorations show an increasing trend in the far future, the mechanical decay will face a general decrease for all types of building materials indoors.

Supplementary Materials: The following are available online at http://www.mdpi.com/2076-3263/8/9/347/s1. The charts of climate change decay variables in relation to the geographical location, time window, material, and size of the building.

Author Contributions: Conceptualization, C.B.; Data curation, A.L.; Methodology, A.L.; Resources, C.B.; Writing—original draft, A.L.; Writing—review & editing, C.B.

Funding: This research received no external funding.

Acknowledgments: This work has been possible thanks to the financial support guaranteed by the Onsager Fellowship-Research Excellence Programme at the Norwegian University of Science and Technology (NTNU) in Trondheim, Norway. The data used in this article has been produced during the Climate for Culture Project within the European Seventh Framework Programme for Research (FP7) under Grant Agreement No. 22 6973.

Conflicts of Interest: The authors declare no conflict of interest.

References

1. Robert, V.; Gobiet, A.; Sobolowski, S.; Kjellström, E.; Stegehuis, A.; Watkiss, P.; Mendlik, T.; Landgren, O.; Nikulin, G.; Teichmann, C. The European climate under a 2 °C global warming. *Environ. Res. Lett.* **2014**, *9*, 034006.

2. Norwegian Climate Service Centre. *Climate in Norway 2100*; Norwegian Climate Service Centre: Oslo, Norway, 2015.

3. Lisø, K.R.; Hygen, H.O.; Kvande, T.; Thue, J.V. Decay potential in wood structures using climate data. *Build. Res. Inf.* **2006**, *34*, 546–551. [CrossRef]

4. Almås, A.J.; Lisø, K.R.; Hygen, H.O.; Øyen, C.F.; Thue, J.V. An approach to impact assessments of buildings in a changing climate. *Build. Res. Inf.* **2011**, *39*, 227–238. [CrossRef]

5. Lisø, K.R.; Hygen, H.O.; Kvande, T.; Thue, J.V.; Harstveit, K. A frost decay exposure index for porous, mineral building materials. *Build. Environ.* **2007**, *42*, 3547–3555. [CrossRef]

6. Larsen, P.K. The salt decay of medieval bricks at a vault in Brarup Church, Denmark. *Environ. Geol.* **2007**, *52*, 375–383. [CrossRef]

7. Grøntoft, F. Climate change impact on building surfaces and façades. *Int. J. Clim. Chang. Strateg. Manag.* **2011**, *3*, 374–385. [CrossRef]

8. Cassar, M.; Young, C.; Weighell, T.; Sheppard, D.; Bomhard, B.; Rosabal, P. *Predicting and Managing the Effects of Climate Change on World Heritage. A Joint Report from the World Heritage Centre, Its Advisory Bodies, and a Broad Group of Experts to the 30th Session of the World Heritage Committee*; World Heritage Committee: Vilnius, Lithuania, 2006.

9. Cassar, M.; Pender, R. The impact of climate change on cultural heritage: Evidence and response. In Proceedings of the ICOM 14th Triennial Meeting, Hague, The Netherlands, 12–16 September 2005; Volume 2, pp. 610–616.

10. Phillips, H. The capacity to adapt to climate change at heritage sites—The development of a conceptual framework. *Environ. Sci. Policy* **2015**, *47*, 118–125. [CrossRef]

11. O'Brien, G.; O'Keefe, P.; Jayawickrama, J.; Jigyasu, R. Developing a model for building resilience to climate risks for cultural heritage. *J. Cult. Herit. Manag. Sustain. Dev.* **2015**, *5*, 99–114. [CrossRef]

12. Gruber, S. The Impact of Climate Change on Cultural Heritage Sites: Environmental Law and Adaptation. *Carbon Clim. Law Rev.* **2011**, *5*, 209–219. [CrossRef]

13. Leissner, J.; Kilian, R.; Kotova, L.; Jacob, D.; Mikolajewicz, U.; Broström, T.; Ashley-Smith, J.; Schellen, H.L.; Martens, M.; van Schijndel, J.; et al. Climate for Culture: Assessing the impact of climate change on the future indoor climate in historic buildings using simulations. *Herit. Sci.* **2015**, *3*, 38. [CrossRef]

14. Sabbioni, C.; Cassar, M.; Brimblecombe, P.; Tidblad, J.; Kozlowski, R.; Drdacky, M.; Saiz-Jimenez, C.; Grontoft, T.; Wainwright, I.; Arino, X. Global climate change impact on built heritage and cultural landscapes. In *Heritage, Weathering and Conservation*; Fort, R., Alvarez de Buergo, M., Gomez-Heras, M., Vazquez-Calvo, C., Eds.; Taylor & Francis Group: Oxford, UK, 2006; pp. 395–401.

15. Nordic Council of Ministers. *CERCMA: Cultural Environment as Resource in Climate Change Mitigation and Adaptation, in Nordiske Arbejdspapirer*; Nordic Council of Ministers: Copenhagen, Denmark, 2014.

16. Kaslegard, A.S. *Climate Change and Cultural Heritage in the Nordic Countries*; Nordic Council of Ministers: Copenhagen, Denmark, 2011.

17. Haugen, A.; Mattsson, J. Preparations for climate change's influences on cultural heritage. *Int. J. Clim. Chang. Strateg. Manag.* **2011**, *3*, 386–401. [CrossRef]

18. Damage Risk Assessment, Economic Impact and Mitigation Strategies for Sustainable Preservation of Cultural Heritage in the Times of Climate Change. Available online: https://cordis.europa.eu/project/rcn/92906_en.html (accessed on 12 September 2018).

19. van Schijndel, A.W.M. *Climate for Culture: Book of Maps*; Eindhoven University of Technology: Eindhoven, The Netherlands, 2016.

20. Jacob, D.; Elizalde, A.; Haensler, A.; Hagemann, S.; Kumar, P.; Podzun, R.; Rechid, D.; Remedio, A.R.; Saeed, F.; Sieck, K.; et al. Assessing the Transferability of the Regional Climate Model REMO to Different COordinated Regional Climate Downscaling EXperiment (CORDEX) Regions. *Atmosphere* **2012**, *3*, 181–199. [CrossRef]

21. Intergovernmental Panel on Climate Change (IPCC). *Special Report on Emissions Scenarios: A Special Report of Working Group III of the Intergovernmental Panel on Climate Change*; Nakicenovic, N., Swart, R., Eds.; Cambridge University Press: Cambridge, UK, 2000.

22. Intergovernmental Panel on Climate Change (IPCC). *Climate Change 2013: The Physical Science Basis. Contribution of Working Group 1 (WG1) to the Fifth Assessment Report of the Intergovernmental Panel on Climate Change*; Stocker, T.F., Qin, D., Plattner, G.K., Tignor, M., Allen, S.K., Boschung, J., Nauels, A., Xia, Y., Bex, V., Midgley, P.M., Eds.; Cambridge University Press: New York, NY, USA, 2013; p. 1535.

23. Huijbregts, Z.; Schellen, H.; Martens, M.; van Schijndel, J. Object Damage Risk Evaluation in the European Project Climate for Culture. *Energy Proced.* **2015**, *78*, 1341–1346. [CrossRef]

24. Ashley-Smith, J. Climate for Culture: D 4.2. Report on Damage Functions in Relation to Climate Change and Microclimatic Response. Available online: https://www.climateforculture.eu/ (accessed on 11 September 2018).

25. Bertolin, C.; Camuffo, D. *Risk Assessment, in Built Cultural Heritage in Times of Climate Change*; Leissner, J., Kaiser, U., Kilian, R., Eds.; Fraunhofer-Center for Central and Eastern Europe MOEZ: Leipzig, Germany, 2014; pp. 52–54.

26. Martens, M.H.J. *Climate Risk Assessment in Museums: Degradation Risks Determined from Temperature and Relative Humidity Data*; Technische Universiteit Eindhoven: Eindhoven, The Netherlands, 2012.

27. Mecklenburg, M.F.; Tumosa, C.S.; Erhardt, D. *Structural Response of Painted Wood Surfaces to Changes in Ambient Relative Humidity in Painted Wood: History and Conservation*; The Getty Conservation Institute: Los Angeles, NY, USA, 1988; pp. 464–483.

28. Jakieła, S.; Bratasz, Ł.; Kozłowski, R. Numerical modelling of moisture movement and related stress field in lime wood subjected to changing climate conditions. *Wood Sci. Technol.* **2008**, *42*, 21–37. [CrossRef]

29. Bratasz, L. Acceptable and non-acceptable microclimate variability: The case of wood. In *Basic Environmental Mechanisms Affecting Cultural Heritage. Understanding Deterioration Mechanisms for Conservation Process*; Camuffo, D., Fassima, V., Havermams, J., Eds.; European Cooperation in Science and Technology: Brussels, Belgium, 2010; Volume 42, pp. 49–58.

30. Grossi, C.M.; Brimblecombe, P.; Menéndez, B.; Benavente, D.; Harris, I.; Déqué, M. Climatology of salt transitions and implications for stone weathering. *Sci. Total Environ.* **2011**, *409*, 2577–2585. [CrossRef] [PubMed]

31. Sabbioni, C.; Brimblecombe, P.; Cassar, M. *The Atlas of Climate Change Impact on European Cultural Heritage: Scientific Analysis and Management Strategies*; Anthem Press: London, UK, 2010.

32. Lankester, P.; Brimblecombe, P. Future thermohygrometric climate within historic houses. *J. Cult. Herit.* **2012**, *13*, 1–6. [CrossRef]

33. Camuffo, D. *Microclimate for Cultural Heritage: Conservation, Restoration, and Maintenance of Indoor and Outdoor Monuments*; Elsevier: New York, NY, USA, 2013.

34. Michalski, S. Double the life for each five-degree drop, more than double the life for each halving of relative humidity. In *Preprints of 13th Meeting of ICOM-CC*; Vontobel, R., Ed.; International Committees Comités Internationaux: London, UK, 2002; Volume 1, pp. 66–72.

35. Sedlbauer, K. *Prediction of Mould Fungus Formation on the Surface of and Inside Building Components*; Fraunhofer Institute for Building Physics: Stuttgart, Germany, 2001.

36. Child, R. The influence of the museum environment in controlling insect pests. In *Climate for Collections: Standards and Uncertainties*; Ashley-Smith, J., Burmester, A., Eibl, M., Eds.; Archetype Publications Ltd.: London, UK, 2013; pp. 419–424.

37. Loli, A.; Bertolin, C. Towards Zero-Emission Refurbishment of Historic Buildings: A Literature Review. *Buildings* **2018**, *8*, 22. [CrossRef]

38. Reinar, D.A.; Westerlind, A.M. *Urban Heritage Analysis: A Handbook about DIVE*; Riksantikvaren: Oslo, Norway, 2010.

39. Lehtimäki, M. *Sustainable Historic Towns—Urban Heritage as an Asset of Development*; National Board of Antiquities: Helsinki, Finland, 2006.

40. Reinar, D.A. *Communicating Heritage in Urban Development Processes 2007–08: Co-Herit Project Report*; Riksantikvaren: Oslo, Norway, 2009.

41. Byplankontoret—Trondheim Kommune, Kulturmiljøet Kjøpmannsgata—En kulturhistorisk Stedsanalyse av Bryggerekken og Tilhørende Områder. Available online: https://www.riksantikvaren.no/ (accessed on 11 September 2018).

42. Valluzzi, M.R.; Modena, C.; de Felice, G. Current practice and open issues in strengthening historical buildings with composites. *Mater. Struct.* **2014**, *47*, 1971–1985. [CrossRef]

43. Modena, C.; Valluzzi, M.R.; da Porto, F.; Casarin, F. Structural Aspects of the Conservation of Historic Masonry Constructions in Seismic Areas: Remedial Measures and Emergency Actions. *Int. J. Archit. Herit.* **2011**, *5*, 539–558. [CrossRef]

44. Cardani, G.; Belluco, P. Reducing the Loss of Built Heritage in Seismic Areas. *Buildings* **2018**, *8*, 19. [CrossRef]

45. Rossi, M.; Cattari, S.; Lagomarsino, S. Performance-based assessment of the Great Mosque of Algiers. *Bullet. Earthq. Eng.* **2015**, *13*, 369–388. [CrossRef]

46. Necevska-Cvetanovska, G.; Apostolska, R. Consolidation, rebuilding and strengthening of St. Clement's church, St. Panteleymon, Plaoshnik, Ohrid. *Eng. Struct.* **2008**, *30*, 2185–2193. [CrossRef]

47. Sesana, E.; Gagnon, A.; Bertolin, C.; Hughes, J. Adapting Cultural Heritage to Climate Change Risks: Perspectives of Cultural Heritage Experts in Europe. *Geosciences* **2018**, *8*, 305. [CrossRef]

48. Hampster Map. Available online: https://www.hamstermap.com/custommap.html (accessed on 11 September 2018).

geosciences

MDPI

Article

Simulations of Moisture Gradients in Wood Subjected to Changes in Relative Humidity and Temperature Due to Climate Change

Charlotta Bylund Melin [1,*], Carl-Eric Hagentoft [2], Kristina Holl [3], Vahid M. Nik [4,5,6] and Ralf Kilian [3]

[1] Department of Preservation and Photography, Nationalmuseum, P.O. Box 16176, SE-10324 Stockholm, Sweden

[2] Department of Architecture and Civil Engineering, Chalmers University of Technology, SE-41296 Gothenburg, Sweden; carl-eric.hagentoft@chalmers.se

[3] Fraunhofer Institute for Building Physics, Fraunhoferstr. 10, 83626 Valley, Germany; kristina.holl@ibp.fraunhofer.de (K.H.); ralf.kilian@ibp.fraunhofer.de (R.K.)

[4] Division of Building Physics, Department of Building and Environmental Technology, Lund University, SE-22363 Lund, Sweden; vahid.nik@byggtek.lth.se or nik.vahid.m@gmail.com

[5] Division of Building Technology, Department of Civil and Environmental Engineering, Chalmers University of Technology, SE-41296 Gothenburg, Sweden

[6] Institute for Future Environments, Queensland University of Technology, Garden Point Campus, 2 George Street, Brisbane 4000, Australia

* Correspondence: charlotta.bylund.melin@nationalmuseum.se; Tel.: +46-70-755-5866

Received: 6 July 2018; Accepted: 10 October 2018; Published: 15 October 2018

Abstract: Climate change is a growing threat to cultural heritage buildings and objects. Objects housed in historic buildings are at risk because the indoor environments in these buildings are difficult to control and often influenced by the outdoor climate. Hygroscopic materials, such as wood, will gain and release moisture during changes in relative humidity and temperature. These changes cause swelling and shrinkage, which may result in permanent damage. To increase the knowledge of climate-induced damage to heritage objects, it is essential to monitor moisture transport in wood. Simulation models need to be developed and improved to predict the influence of climate change. In a previous work, relative humidity and temperature was monitored at different depths inside wooden samples subjected to fluctuating climate over time. In this article, two methods, the hygrothermal building simulation software WUFI® Pro and the Simplified model, were compared in relation to the measured data. The conclusion was that both methods can simulate moisture diffusion and transport in wooden object with a sufficient accuracy. Using the two methods for predicted climate change data show that the mean RH inside wood is rather constant, but the RH minimum and maximum vary with the predicted scenario and the type of building used for the simulation.

Keywords: moisture transport; wood; relative humidity; climate variations; measurements; experimental research; hygrothermal simulation models; typical and extreme weather conditions; climate change

1. Introduction

The growing threat of climate change to cultural heritage has gained increasing awareness. The knowledge is due to findings from research projects which aim to predict future climate change and its impact on cultural heritage buildings and the indoor environment in those buildings. The Global Climate Change Impact on Built Heritage and Cultural Landscapes; Noah's Ark Project (2004–2007) brought forward the fact that little attention had been paid to the impact of global climate change on cultural heritage and that this needed to be better recognised and perceived as relevant. Due to climate change, a range of direct and indirect effects were expected to be observed on built heritage [1,2]. The EU research project Climate for Culture: Damage Risk Assessment, Economic Impact and Mitigation Strategies for Sustainable Preservation of Cultural Heritage in Times of Climate Change (2009–2014) studied the impact and mitigation strategies for preservation of cultural heritage in times of climate change. The project developed simulation models to estimate the impact of future global climate change on the indoor environments in different types of buildings in different regions of Europe. According to the project, the indoor temperature (T) in non-heated buildings in parts of northern Europe will at first (2021 to 2050) increase, but in the far future (2071 to 2100) decrease [3]. Important research projects on improving energy efficiency of historic built heritage include, for instance, Sustainable Energy Communities in Historic Urban Areas (SECHURBA) (2008–2011) [4], and Energy Efficiency for EU Historic Districts' Sustainability (EFFESUS) (2012–2016) [5].

So far, the main focus on energy efficiency measures of these projects has been on the buildings and the indoor environment and less on the effect on the objects housed in them. Historic buildings, such as churches, often have large interior volumes and a high air infiltration rate, which obstructs efforts to regulate indoor relative humidity (RH) and T. The buildings themselves are often of high cultural heritage value. Therefore, interventions, such as installation of air conditioning plants or alterations to the building envelope to decrease the air infiltration, are often restricted. Consequently, the interiors may be subjected to large daily, as well as seasonal, changes in both RH and T, much larger than the recommended climate criteria for hygroscopic museum objects [6]. For these reasons, it is important to also include hygrothermal monitoring and monitoring of mechanical deformation of heritage objects located in historic buildings. Moreover, it is central to find reliable modelling methods to be able to predict potential future impact of climate change. This was recognised by the Netherlands Organisation for Scientific Research (NWO) and Rijksmuseum Amsterdam in their report, The Conservation of Panel Paintings and Related Objects: Research agenda 2014–2020. It emphasised that a balance between preservation of art, energy cost and effects on buildings in the widest sense should be encouraged. It further suggests research topics which should comprise: modelling behaviour patterns including validation studies, experimental population studies, hygro-mechanical properties of ageing wood in panels and inter-laminar stress and fracture mechanics, which also affect paint layers [7].

Hygroscopic organic materials, such as wood, are particularly susceptive to changes in the ambient climate. With an increase in RH, wood will adsorb moisture from the ambient air and swell. With a decrease in RH, it will desorb moisture and shrink. If the changes in RH and T are significant, or frequent enough, permanent deformation or damage may occur. The moisture content (MC) in wood is defined as the mass of water in relation to the oven-dried wood, expressed as a percentage. Maximum swelling or shrinkage at certain RH may occur when equilibrium moisture content (EMC) is reached. It is defined as the MC at which the wood is neither adsorbing nor desorbing moisture from the ambient air. However, equilibrium will only follow if RH and T are constant for a long enough period of time for the wood to be fully acclimatised to the ambient air throughout. This may take a very long time and during real-life conditions it is uncertain if EMC is ever reached [8]. In a fluctuating climate, constantly moving moisture gradients will develop from the surface and inwards. Methods which can accurately monitor moisture movement in wood due to different RH and T combinations are few. Nevertheless, the study of moisture diffusion in wood is an important first step since it will contribute to an increased understanding of how deformation of wood and, thereby, also damage, or lack of damage, develop [9]. Various types of long-term wood electrical resistance sensors to monitor MC to predict the service life of wooden constructions have been tested [10–13]. However, some resistance methods have shown to be connected with measuring errors. Because wood shrinks and swells, the contact between the wood and the resistance pins may vary, resulting in inaccurate readings [11]. The volume of the wooden sample and the distance of sensor to the surface do also affect the measuring results [14]. Therefore, a method to monitor RH and T distribution in wooden samples was instead developed, and is described in [15]. The monitored data can also easily be converted to MC according to Equation (2) of this article. The method is based on small RH and T sensors, which are inserted into drilled holes in wooden samples. The sensors are located at different depths in order to monitor the moisture movement inside the wood. The samples were exposed to step-changes and fluctuations in RH and T in a climate chamber. The advantage is that the method can be used in in situ monitoring campaigns in historic buildings [16]. These data can be further used for simulation modelling to predict, for instance the future effects of climate change on wooden objects. To measure RH and T inside the material instead of using the general room climate data reduces the risk of misinterpretation since local microclimates found in historic buildings, for instance behind paintings hanging on walls or inside closed cabinets, are also influencing the objects.

To study the effect of climate change on heritage objects housed indoors, it is important to take the building type into account, because their response to the outdoor climate will influence the indoor climate. Climate change induces variations in both long- and short-term behaviour of the climate system, resulting in warmer weather with stronger and more frequent extreme conditions [17]. Such variations can affect the hygrothermal performance of building components on different time scales [18,19]. Hence, the actual effect on the objects is the combination of the outdoor climate and the kind of building.

In the first part of this article, the aim was to further develop and compare simulation methods which can study the hygrothermal effect on wooden objects during variations in RH and T. It is based on existing data from previously performed laboratory experiments [16]. Two simulation methods were chosen: WUFI® Pro software and a simplified analysis method (Simplified model). In the second part, the aim was to simulate the effect of predicted climate change scenarios to wooden objects in different types of buildings. This was done by using two different methods. Firstly, a verified method used for the impact assessment of climate change, which is based on synthesizing three sets of 1-year weather data sets, representing typical, extreme-cold and extreme-warm conditions [20,21]. The method helps to run faster simulations while climate uncertainties and extreme conditions are taken into account. The synthesized representative weather data sets were then applied to two different types of generic buildings, i.e., a typical heavy and a typical light constructed house. Secondly, within the Climate for Culture project, simulated climate data due to climate change has also been produced [22,23]. These data were used by the Fraunhofer Institute for building physics in order to generate data of the indoor climate for certain case studies all over Europe. To show the simulated effect of future climate change on the indoor climate an existing building, Roggersdorf church in Bavaria, Germany, was also used in this study. It is a small heavyweight building without any strategy for climate control. The generated climate data was finally used to study the simulated effect on the moisture distribution in wood assumed housed in the generic and existing buildings according to the WUFI® Pro model.

The principal conclusions are that both WUFI® software program and the Simplified methods are capable of quite accurate predictions of the moisture conditions inside wood at temperature and RH variations of 7–25 °C to 35–75%. Likewise, the two methods are generally in agreement, while the influence on wood is generated from simulated climate change data. The mean values are rather constant during the simulated periods. However, different types of climate predictions from different kinds of buildings generate variances in minimum and maximum RH inside wood.

2. Materials and Methods

2.1. Experimental Data

In a previous work by Bylund Melin et al., a method was developed and thoroughly presented to monitor moisture transfer over time. Tangential cut wood samples (Scots pine, *Pinus sylvestris* L.) were subjected to various changes in ambient climate in a laboratory climate chamber. The aim of this work was to study the impact of the effect of fluctuating RH and T which can be found in less climate-controlled historic buildings [15]. In this study, Scots pine was chosen, since it is a

Geosciences **2018**, *8*, 378

wood species common in cultural heritage objects in Sweden. The method is also designed to be used with other types of wood. The dimensions of the wood samples used in the experiments were 200 mm × 45 mm × 45 mm. The monitoring device consists of small RH and T sensors (MSR loggers), which were inserted from the reverse side of the wooden samples down to different depths (1, 4 and 7 mm from the front side). Due to the monitoring at several depths, it is possible to study in detail the effects of changing RH or T as well as the combined effect of changing RH and T. The method can be used in controlled laboratory settings as well as in situ locations, such as historic buildings. The chosen dataset used here was taken from [16]. It consists of 10-day long step-changes, in which the ambient RH in the ambient climate chamber varied between 35 and 75% RH. The measured data used during the period from day 40 to day 60 consists of two measuring periods, which were attached [16] (Figure 2). Less climate-controlled historic buildings can suffer from RH well above 75%, and heated buildings often show very low RH levels during winter and should have been included in the experiments. However, due to limitations of the climate chamber at low temperatures, the experiments were limited to the 35–75% range [16].

The monitored and simulated data used in this article are publicly available at Chalmers University of Technology: http://www.byggnadsteknologi.se/.

2.2. Simplified Theoretical Analysis

Several numerical simulation programs were analysed in the HAMSTAD project [24]. The project presented five numerical benchmark cases for the quality assessment of simulation models for one-dimensional heat, air and moisture (HAM-) transfer. Several solutions from different universities and institutes were compared. Consensus solutions could be found. However, the various presented calculation results varied somewhat. The Simplified model is based on a linearization of the sorption curve and constant water vapour diffusion. It does not need a specific software program; instead the calculations can be performed in a simple Excel spreadsheet. In [24], various types of numerical solutions are applied to handle the moisture transfer benchmark cases. These give a background to the complexity of the problem at hand and the expected acceptable accuracy.

The Simplified model is presented in this section. In this, hysteresis is neglected and the sorption isotherm is assumed to be linear. The moisture transfer is driven by the gradient in humidity by volume, v (kg/m^3). The transport coefficient, δ_v^0 (m^2/s), is assumed to be constant, i.e., independent of moisture levels. The developed model allows for the development of handy analytical expressions which can give a lot of insights. In Section 4.1, it is also shown that the simplified analysis gives reasonable results in a comparison with experimental results.

The moisture balance equation assuming constant, but time dependent, temperature through the material becomes, with $\varphi(x,t)$ (−) representing the relative humidity in the material:

$$-\frac{\partial}{\partial x}\left(-\delta_v^0 \frac{\partial v}{\partial x}\right) = \frac{\partial w}{\partial t}$$

$$\Leftrightarrow$$

$$\delta_v^0 \frac{\partial^2 v}{\partial x^2} = \delta_v^0 v_s(T)\frac{\partial^2 \varphi}{\partial x^2} = \frac{\partial w}{\partial t} = \frac{\partial w}{\partial \varphi}\frac{\partial \varphi}{\partial t} = \zeta \frac{\partial \varphi}{\partial t} \tag{1}$$

Here, $w(\varphi)$ (kg/m^3) is the moisture content per volume unit, which depends on the relative humidity only since hysteresis is neglected in the simplified model. The relation $v = v_s(T) \cdot \varphi$, where v_s represents the humidity by volume at saturation has been used. The following relation with the moisture content $u(-)$ (sometimes referred to as the MC) can be used to translate between units:

$$u = \frac{w(\varphi)}{\rho_{dry}} \tag{2}$$

The term in the denominator is the dry density of the wood (kg/m^3).

As a part of the simplified model, the slope of the sorption curve is assumed to be constant:

$$\frac{\partial w}{\partial \varphi} = \xi \qquad (3)$$

Introducing the water vapour moisture diffusivity a_v (m^2/s):

$$\frac{\partial^2 \varphi}{\partial x^2} = \frac{1}{a_v(t)} \frac{\partial \varphi}{\partial t} \qquad a_v(t) = \frac{\delta_v^0 v_s(T(t))}{\xi} \qquad (4)$$

The next step is to analyse the more complicated case with simultaneous step-wise variations in RH and boundary temperature. The wood panel is exposed to the following varying load at $x = 0$ neglecting any surface resistances:

$$\begin{cases} \varphi_i = \varphi_0 + \sum_{n=1}^{N} (\varphi_n - \varphi_{n-1}) \cdot H(t - t_n) \\ T_i = T_0 + \sum_{n=1}^{N} (T_n - T_{n-1}) \cdot H(t - t_n) \end{cases} \qquad t > 0 \quad t_n > 0 \qquad (5)$$

Here, $H(t)$ represents the Heaviside unit-step function; it is equal to zero for times less than zero, and one for times greater than zero.

Before time t_1 the wood sample has been exposed to a stable climate (RH, temperature) of (φ_0, T_0) for a very long time. The step at time t_n change the boundary value both for the relative humidity and the temperature by the amount $(\varphi_n - \varphi_{n-1})$ and $(T_n - T_{n-1})$ respectively. Equation (4) needs to be solved with boundary condition (5).

Simplified Analysis—Step-Change, Periodic Variation and Time Varying Moisture Diffusivity

First some simple, but handy, analytical solutions for cases from [25] with constant temperature, $T = T_0$, and semi-infinite region will be presented.

With one step change at time zero at the wood surface at $x = 0$:

$$\varphi_i = \varphi_0 + \Delta\varphi \cdot H(t) \quad t > 0 \qquad (6)$$

The analytical solution [25] for a semi-infinite domain, with constant temperature, is:

$$\varphi(x, t) = \varphi_0 + \Delta\varphi \cdot \mathrm{erfc}\left(\frac{x}{\sqrt{4a_v \cdot t}}\right) \qquad x \geq 0, t \geq 0 \qquad (7)$$

Here, erfc is the complimentary error function.

The penetration depth, i.e., the depth to which approximately half the disturbance of what happened at the boundary has propagated:

$$x_{0.5} = \sqrt{a_v \cdot t} \qquad (8)$$

Typically, for wood, with $\delta_v^0 = 0.5 \times 10^{-6}$ m^2/s and $\xi = 120$ kg/m^3 at 21 °C [25], this depth is around 0.0005 m after 1 h, 0.0007 m after 2 h, 0.003 m after 1 day and 0.007 m after a week. These very limited penetration depths also mean that the assumption of semi-infinite region is not really a limitation. This assumption is valid if the real thickness of the material layer is on the order of 2–5 times the penetration depth.

The following variation at the wood surface at $x = 0$ is given:

$$\varphi_i = \varphi_0 + \varphi_A \cdot \sin\left(\frac{2\pi t}{t_p}\right) \qquad (9)$$

Here, t_p (s) is the time period of the sinusoidal variation. The analytical solution [25] becomes:

$$\varphi(x,t) = \varphi_0 + \varphi_A \cdot e^{-x/d_{pv}} \sin\left(\frac{2\pi t}{t_p} - x/d_{pv}\right) \qquad x \geq 0 \tag{10}$$

The penetration depth d_{pv} (m), i.e., the depth were the amplitude of the RH has diminished with a factor $\exp(-1)$, approx. 0.37 reads:

$$d_{pv} = \sqrt{\frac{a_v t_p}{\pi}} \tag{11}$$

Typically, for wood, with $\delta_v^0 = 0.5 \times 10^{-6}$ m^2/s and $\xi = 120$ kg/m^3 at 21 °C [25], this depth is around 0.001–0.002 m for a diurnal variation ($t_p = 24$ h) and 0.03 m for a yearly one.

The following variable substitution is introduced in order to solve (4) with time-dependent moisture diffusivity:

$$\tau(t) = \int_0^t a_v(t')\, dt' \tag{12}$$

This changes (4) to:

$$\frac{\partial^2 \varphi}{\partial x^2} = \frac{\partial \varphi}{\partial \tau} \tag{13}$$

This equation is similar to the classic one-dimensional heat conduction or diffusion equation with the diffusivity term equal to one. The equation is linear when using this transformed time variable; thus, superposition techniques can be used. Therefore, only the solution of a unit-step change is needed to handle the boundary conditions according to (5).

It is assumed that the penetration of moisture into the wood is much less than the thickness of the sample and that the initial relative humidity is ϕ_0. With a step in relative humidity of $\Delta\phi$, we then get:

$$\varphi(x,\tau) = \varphi_0 + \Delta\varphi \cdot \text{erfc}\left(\frac{x}{\sqrt{4\tau}}\right) \qquad x \geq 0, \tau \geq 0 \tag{14}$$

The complete solution, referred to as the Simplified model, of (4) and (5) then becomes:

$$\varphi(x,t) = \varphi_0 + \sum_{n=1}^{N} (\varphi_n - \varphi_{n-1}) \cdot erfc\left(\frac{x}{\sqrt{4 \cdot \tau(t - t_n)}}\right) \tag{15}$$

2.3. WUFI® Pro Simulation Method

WUFI® is a well-known method to calculate transient heat and moisture transport in building materials. It was created by Künzel, who developed a differential equation system based on the physical principles of heat and moisture transport for determining the moisture behaviour of multilayer building structures under natural climatic boundary conditions [26]. This was numerically implemented in the WUFI® software program, further refined at the Fraunhofer Institute for Building Physics and verified with the assistance of the experimental field test site in Holzkirchen. The program WUFI® Pro is suitable to simulate the temperature and moisture transport inside individual layers of composite materials. WUFI® Pro was originally developed for the simulation of the hygrothermal behaviour of construction parts. The program was first used for the examination of the behaviour of artworks when exposed to climatic fluctuations by Holl. The simulations were validated by a dummy painting on canvas with determined material data, which was put with different distances on the inside of an exterior wall of a test building [27].

Since the WUFI® material database does not include data for Scots pine, the simulations were instead carried out on Spruce (radial cut). However, in contrast with the Simplified model, the database includes data on the sorption isotherm as well as the water vapour diffusion resistance. The value for water vapour diffusion equivalent air layer thickness (sd-value) for the reverse side of the wooden samples was set at 1000 m, since the wooden samples were actually covered with aluminium foil on all but one side (the measuring front side). The initial RH was 77% and the initial T was 15.4 °C.

3. Impact of Future Climate Change on Wood

The Simplified model and WUFI® Pro simulation method were also used to study the impact of predicted climate change on wooden objects housed in two different generic buildings located in Gothenburg, Sweden and an existing church (Roggersdorf church) in Bavaria, Germany. Two different hygrothermal simulation methods were used to predict the climate change and are presented below.

3.1. Hygrothermal Simulations of Future Climatic Conditions in Two Generic Buildings

Due to the existence of climate uncertainties and the need for considering several future climate scenarios, there will be large datasets to take into account, which makes the assessment time-consuming [28]. A method has been developed for creating representative weather data sets for future climatic conditions, considering typical and extreme conditions [20]. More than energy simulations, the proposed approach has been tested and verified for the hygrothermal simulation of building components by simulating the moisture conditions in the outer façade layer of a wooden frame wall in WUFI [21]. The approach was adopted in this work, which is based on synthesizing and using three sets of 1-year weather data, representing future climatic conditions for 2070–2099: (1) typical downscaled year (TDY), (2) extreme cold year (ECY), and (3) extreme warm year (EWY). The representative weather data were synthesized out of RCA4, the 4th generation of the Rossby Centre regional climate model (RCM) [29]. Considering Gothenburg and two Representative Concentration Pathways (RCPs) in this study, RCA4 downscaled three global climate models (GCMs) to the spatial resolution of 12.5 km^2: CNRM-CM5 (for RCP4.5 and RCP8.5), ICHEC-EC-EARTH (for RCP4.5 and RCP8.5) and IPSL-CM5A-MR (for RCP8.5), resulting in five different climate scenarios. This means that the 1-year representative weather data sets (TDY, ECY and EWY) were synthesized considering 30 years of data for five different scenarios. More details on climate scenarios and calculating climate parameters are available in [20,30].

The three sets of 1-year weather data were applied on two types of generic buildings: (1) A non-habited light house (vapour concentration indoors equal to the outdoors and indoor T based on floating 24 h value of the outdoor T, plus 2 °C) and a heavy house (vapour concentration indoors equal to outdoors, plus 0.5 g/m^3, indoor T based on floating one week value of the outdoor T plus 1 °C).

3.2. Hygrothermal Simulations of Future Climate Conditions in Roggerdorf Church

Within the Climate for Culture project, the Max Planck institute produced data sets for the recent past (1960–1990), the near future (2020–2050) and the far future (2070–2100) for the calculated future climate scenario A1B [31]. These three scenarios were applied to the Roggersdorf church by the Fraunhofer Institute for building physics in order to generate data of the indoor climate and the effect on wood.

4. Results

4.1. Comparison of the WUFI® Pro and the Simplified Model

The measurements performed in [16] were used to validate the Simplified model and to compare it with WUFI® Pro and the measured data. In these measurements the boundary RH and temperature vary in intervals of 10 days over a whole period of 100 days. In the Simplified model, this means that a_v basically varies every 10-day period. Thus, the variable $\tau(t)$ (12) is represented by a continuous curve built up by piece-wise linear segments, with the slope depending on a_v. The data used in the simulation was $\delta_v = 0.5 \times 10^{-6}$ m^2/s and $\xi = 80$ kg/m^3.

The results for WUFI® Pro and the Simplified model in relation to the measured data at the depth of 1, 4 and 7 mm can be seen in Figures 1–3. According to the error analysis and comparison with the HAMSTAD benchmark, both simulation methods show generally excellent compliance with the measured data. At 1 mm depth (Figure 1), the two simulation methods overestimate the results on desorption, while on adsorption they are much closer to the measured values. The WUFI® calculation method is generally closer to the measurements than the Simplified model. This tendency is not as clear at 4 and 7 mm depth. In fact, on 7 mm (Figure 3) the simulation on desorption is more conformed compared to the adsorption.

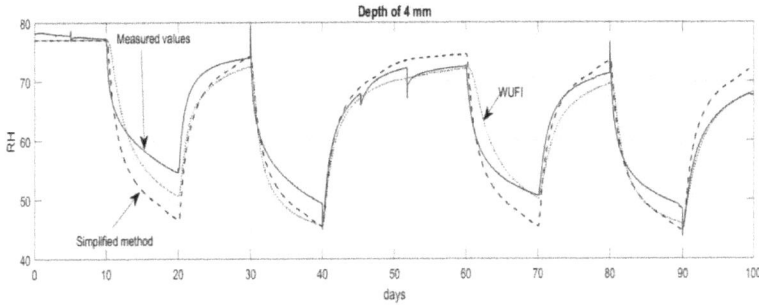

Figure 1. The results at 1 mm depth of the calculated RH in comparison with the measured data. The blue solid line indicates the measured data, the black dotted line is the simulation by WUFI® Pro and the red dashed line the Simplified model.

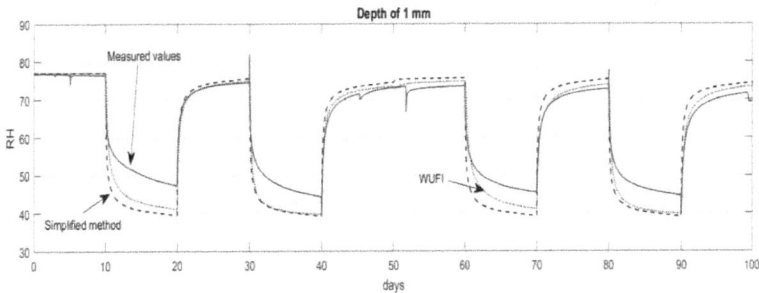

Figure 2. The results at 4 mm depth of the calculated RH in comparison with the measured data. The blue solid line indicates the measured data, the black dotted line is the simulation by WUFI® Pro and the red dashed line the Simplified model.

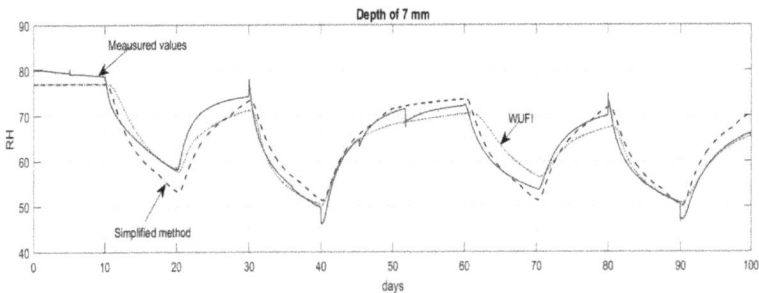

Figure 3. The results at 7 mm depth of the calculated RH in comparison with the measured data. The blue solid line indicates the measured data, the black dotted line is the simulation by WUFI® Pro and the red dashed line the Simplified model.

In summary, Table 1 shows the average difference and standard deviation between the two models and the measured data. The two methods are similar in their results, and both methods show a larger difference at 1 mm depth than at 7 mm depth.

Table 1. Average difference and standard deviation based on hourly values between measured and modelled result in the 100 days measured.

	The Simplified Model		WUFI® Pro Simulations	
Depth (mm)	Average Difference (%)	Standard Deviation (%)	Average Difference (%)	Standard Deviation (%)
1 mm	1.6	4.9	1.6	3.4
4 mm	1.1	3.2	1.2	2.5
7 mm	0.1	2.9	0.3	2.5

Despite the reported average difference and rather small standard deviation (Table 1), the WUFI® calculation method shows a generally closer agreement with the measured data compared to the Simplified model. The results are not in full agreement throughout, as can be seen in Figures 1–3.

4.2. The Effect on Wood Using Hygrothermal Simulations of Future Climatic Conditions

The hygrothermal influence at 1, 4 and 7 mm depth in wood due to predicted climate change are presented in Figures 4–6. For the Roggersdorf church, simulation only the WUFI® Pro model was used.

Distributions of the calculated RH values inside the wood at different layers are shown for the light and heavy buildings respectively in box-and-whiskers plots in Figures 4 and 5. Distributions are divided into four major groups; the first three (TDY, ECY and EWY) are based on the applied weather data (see Section 3.1) and the last one (Triple) contains all three groups. As has been shown previously [20,21], the distribution of typical and extreme conditions together is the most representative one.

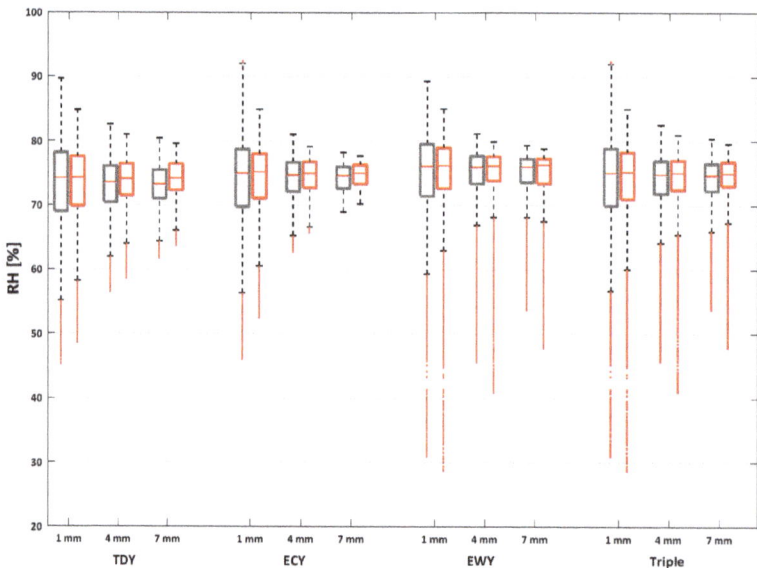

Figure 4. RH distribution in wood at different depths, using the WUFI® Pro method (red boxes) and the Simplified model (grey boxes). Results are for the generic light building (1 year data) subjected to three weather data sets; typical downscaled year (TDY), extreme cold year (ECY) and extreme warm year (EWY). Triple set represents distribution of all the three data sets together.

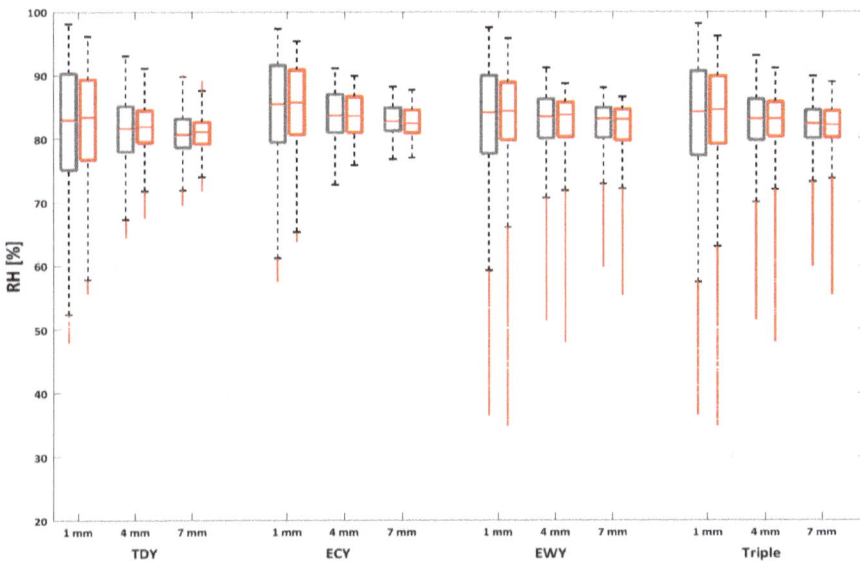

Figure 5. RH distribution in wood at different depths, using WUFI® Pro (red boxes) and the Simplified model (grey boxes). Results are for the typical heavy building subjected to three weather data sets; typical downscaled year (TDY), extreme cold year (ECY) and extreme warm year (EWY). Triple set represents distribution of all the three data sets together.

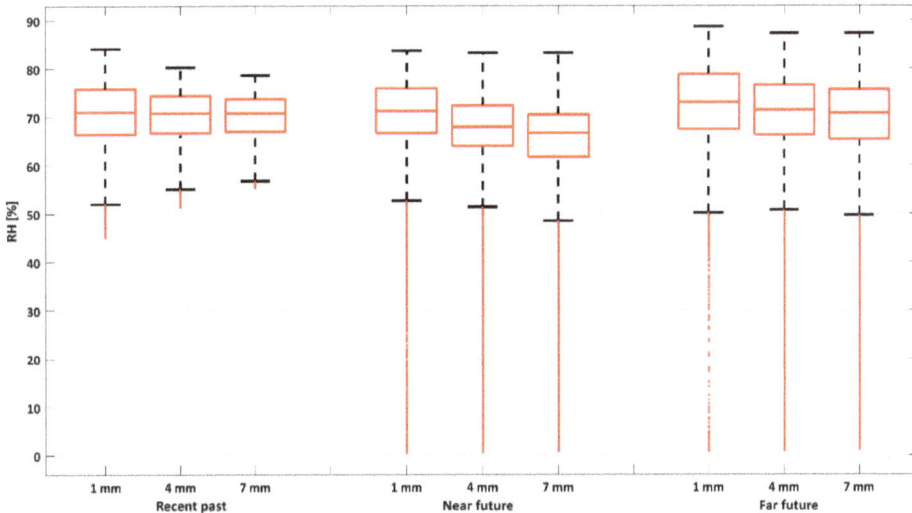

Figure 6. RH distribution in wood at different depths, using the WUFI® Pro method using the indoor data from Roggersdorf church for three 30-year time periods: recent past (1960–1990), near future (2020–2050) and far future (2070–2100).

Comparing the WUFI® Pro method and the Simplified model shows that the mean values indicate close correlation. However, the Simplified model results in wider distributions of RH values, which is visible by having larger interquartile ranges (compare the size of boxes between two methods in the figures) and whiskers (outliers are almost in the same range for both methods, highly influenced by weather conditions). By getting deeper in the wood, differences between two methods decrease. For example, RH distributions among two methods are more similar in the depth of 7 mm than 1 mm.

For the generic buildings, it is clear that the mean RH in wood is lower in the light building type (approximately 73–77% RH at all depths) and higher in the heavy building type (approximately 80–85% RH at all depths). However, RH is generally more stable in the wood located in the heavy building, which is shown by the size of the larger boxes and the more bunched whiskers and outliers. An extreme cold year results in more stable RH inside the wood compared to an extreme warm year.

The data from Roggersdorf church (Figure 6) shows mean values of approximately 67–73% RH, slightly lower than the light building in Figure 4. RH is reduced in the near future, but increases in the far future. The distribution of RH inside wood is much larger (smaller boxes and more spread out whiskers) for Roggersdorf in relation to the two generic examples.

The predicted outdoor data from the two regions where the buildings are located are shown in Table 2. It shows that RH is higher in Gothenburg, located on the west coast of Sweden in comparison with Bavaria, which is located inland. However, average RH is similar within each prediction situation (82.98–84.15% RH in Gothenburg and 70.4–72.1% RH in Bavaria). The temperature increases in Bavaria but does not reach the extreme warm weather predicted in Gothenburg. Due to the forecasted increase in T and RH for the far future scenario in Roggersdorf, T and RH inside the wooden samples in the Roggersdorf church increase as well (Figure 6). Due to the different time scales of the two prediction models, further comparison is difficult.

Table 2. Predicted yearly average outdoor T and RH in Gothenburg (affecting the generic buildings) and the 30 years average of Bavaria (affecting the Roggersdorf church).

	Gothenburg (One Year)			Bavaria (30 Years)	
	Temperature (°C)	Relative Humidity (%)		Temperature (°C)	Relative Humidity (%)
TDY	9.109	82.98	Recent past	9.8	70.4
ECY	4.131	85.08	Near future	10.6	71.2
EWY	13.54	84.15	Far future	11.6	72.1

5. Discussion

To monitor moisture transport in wood (and other cultural heritage materials), it is essential to be able to validate and adjust simulation methods. This is the first effort known to the authors and it gave unexpectedly good results. It is believed that both methods can be used and developed further to estimate the impact of altered indoor environments to wooden objects in historic buildings subjected to changing heating regimes of the buildings or due to global warming. Although the WUFI® Pro simulation performed slightly better, the Simplified model has an advantage in that it is easy to use and does not need specific software. It is assumed that for the simulation using a complex composite material such as a panel painting, WUFI® will be more accurate, but this still has to be proven. The previously mentioned HAMSTAD project [24] presents a spread in results between different numerical methods. Even though the benchmark cases were not the same as the one in this paper, the performance of the Simplified method can very well match any of the other numerical ones, i.e., the difference between the Simplified model and WUFI® Pro is in the parity of the difference between the different numerical solutions in [24]. The Simplified model presented assumes a semi-infinite flow domain. This may sound limiting but the method is applicable with good accuracy as long as the penetration depth is smaller or of the same magnitude as the exposed layer thickness. The penetration depth is on the order of millimetres, as shown in Section 2.2. The Simplified model can rather easily be extended to also cover the case with surface resistances.

To the best knowledge of the authors, this is the first time a simple analytical solution for the penetration of moisture in to wood during cycling of both temperature and RH is presented that can match a state-of-the-art numerical moisture transfer program such as WUFI® Pro in accuracy. The model can rather easily be incorporated in a simple spreadsheet program such as Excel to calculate durability indicators.

Using the two methods to study the impact of climate change was tested within this work. Some of the results were expected, for instance that RH in the wood was higher during extreme cold weather conditions compared to extreme warm weather conditions. The larger difference between minimum and maximum RH inside the wood during summer could result in larger mechanical strain of wood and consequently permanent deformation. On the other hand, during extreme cold, the generally higher RH can result in increased risk for mould growth.

Interestingly, by getting deeper in the wood, differences in RH between the two methods decrease. This might be because of a difference in the boundary condition. In WUFI® Pro, a water vapour surface resistance is considered, while this is omitted in the Simplified model. This difference is of minor importance deeper into the wood. Another explanation could be the depth of the drilled holes. Especially at 1 mm depth, there is a risk that the thickness is not exact, and this can make the results uncertain. It is possible that this is one reason there is a larger difference between the measures and simulated data at this depth compared to 7 mm. Therefore, the method used here should be validated further.

The use of spruce instead of Scots pine in the WUFI® Pro hygrothermal simulation software poses an uncertainty in the simulation. To be more precise in the simulation, it is necessary to be as realistic as possible in the input data.

6. Conclusions

It has been shown that both methods, the Simplified model and the WUFI® Pro hygrothermal simulation software, are able to simulate moisture diffusion and transport in wooden objects with sufficient accuracy. Using them to predict the effect of future climate change gave likely results, which further validate the methods. It gives a good indication of how wooden objects will react due to future climate change. Several future studies are possible while further developing the measuring method and refining the models. An example would be monitoring moisture transport in three-dimensional objects subjected to the environment from more than one side as well as painted wooden objects. To relate moisture transport to deformation (elastic and plastic) of wooden samples and real objects is also an important future task in order to better understand and assess climate-related damage processes to valuable cultural heritage artefacts.

Author Contributions: C.B.M. contributed with the experimental data. C.-E.H. conceived the Simplified model and Kristina Holl the WUFI® Pro simulations on the experimental data. V.M.N. synthesized future weather data and assessed the results. C.B.M., C.-E.H., K.H., R.K. and V.M.N. wrote equal parts of the article.

Acknowledgments: The authors wish to thank Nayoka Martinez-Bäckström for editing and proof-reading the English language of this article.

Conflicts of Interest: The authors declare no conflict of interest.

References

1. Sabbioni, C. Noahs Ark Report Summary: Final Report Summary—Noahs Ark (Global Climate Change Impact on Built Heritage and Cultural Landscapes). 2011. Available online: https://cordis.europa.eu/result/rcn/47770_en.html (accessed on 8 April 2017).
2. Anthem Press. *The Atlas of Climate Change Impact on European Cultural Heritage: Scientific Analysis and Management Strategies*; Sabbioni, C., Brimblecombe, P., Cassar, M., Eds.; Anthem Press: London, UK, 2012.
3. Leissner, J.; Kilian, R.; Kotova, L.; Jacob, D.; Mikolajewicz, U.; Broström, T.; Ashley-Smith, J.; Schellen, H.L.; Martens, M.; van Schijndel, J.; et al. Climate for Culture: Assessing the impact of climate change on the future indoor climate in historic buildings using simulations. *Herit. Sci.* **2015**, *3*, 1–15. [CrossRef]
4. Sustainable Energy Communities in Historic URBan Areas (SECHURBA). Available online: https://ec.europa.eu/energy/intelligent/projects/en/projects/sechurba#partners (accessed on 27 July 2018).
5. EFFESUS. Available online: http://www.effesus.eu/ (accessed on 27 July 2018).

6. Environmental Guidelines: IIC and ICOM-CC Declaration. 2014. Available online: https://www.iiconservation.org/node/5168 (accessed on 10 November 2014).
7. Netherlands Organisation for Scientific Research. *The Conservation of Panel Paintings and Related Objects: Research Agenda 2014–2020*; Kos, N., van Duin, P., Eds.; Netherlands Organisation for Scientific Research (NWO): The Hague, The Netherlands, 2014. Available online: https://rkd.nl/en/explore/library/288805 (accessed on 13 June 2017).
8. Engelund, E.T.; Garbrecht Thygesen, L.; Svensson, S.; Hill, C.A.S. A critical discussion of the physics of wood–water interactions. *Wood Sci. Technol.* **2013**, *47*, 141–161. [CrossRef]
9. Bylund Melin, C. Wooden Objects in Historic Buildings: Effects of Dynamic Relative Humidity and Temperature. Ph.D. Thesis, University of Gothenburg, Gothenburg, Sweden, January 2018.
10. Dai, G.; Ahmet, K. Long-term monitoring of timber moisture content below the fiber saturation point using wood resistance sensors. *For. Prod. J.* **2001**, *51*, 52–58.
11. Brischke, C.; Rapp, A.O.; Bayerbach, R. Measurement system for long-term recording of wood moisture content with internal conductively glued electrodes. *Build. Environ.* **2008**, *43*, 1566–1574. [CrossRef]
12. Fredriksson, M.; Wadsö, L.; Johansson, P. Small resistive wood moisture sensors: A method for moisture content determination in wood structures. *Eur. J. Wood Wood Prod.* **2013**, *71*, 515–524. [CrossRef]
13. Isaksson, T.; Thelandersson, S. Experimental investigation on the effect of detail design on wood moisture content in outdoor above ground applications. *Build. Environ.* **2013**, *59*, 239–249. [CrossRef]
14. Fredriksson, M.; Claesson, J.; Wadsö, L. The Influence of specimen size and distance to a surface on resistive moisture content measurements in wood. *Math. Probl. Eng.* **2015**, *2015*, 1–7. [CrossRef]
15. Bylund Melin, C.; Gebäck, T.; Heintz, A.; Bjurman, J. Monitoring dynamic moisture gradients in wood using inserted relative humidity and temperature sensors. *E-Preserv. Sci.* **2016**, *13*, 7–14. Available online: http://www.morana-rtd.com/e-preservationscience/2016/ePS_2016_a2_Bylund_Melin.pdf (accessed on 17 March 2018).
16. Bylund Melin, C.; Bjurman, J. Moisture gradients in wood subjected to RH and temperatures simulating indoor climate variations as found in museums and historic buildings. *J. Cult. Herit.* **2017**, *25*, 157–162. [CrossRef]
17. IPCC. *Climate Change 2013: The Physical Science Basis. Contribution of Working Group I to the Fifth Assessment Report of the Intergovernmental Panel on Climate Change*; Cambridge University Press: Cambridge, UK; New York, NY, USA, 2013.
18. Nik, V.M.; Kalagasidis, A.S.; Kjellström, E. Statistical methods for assessing and analysing the building performance in respect to the future climate. *Build. Environ.* **2012**, *53*, 107–118. [CrossRef]
19. Nik, V.M.; Kalagasidis, A.S.; Kjellström, E. Assessment of hygrothermal performance and mould growth risk in ventilated attics in respect to possible climate changes in Sweden. *Build. Environ.* **2012**, *55*, 96–109. [CrossRef]
20. Nik, V.M. Making energy simulation easier for future climate—Synthesizing typical and extreme weather data sets out of regional climate models (RCMs). *Appl. Energy* **2016**, *177*, 204–226. [CrossRef]
21. Nik, V.M. Application of typical and extreme weather data sets in the hygrothermal simulation of building components for future climate—A case study for a wooden frame wall. *Energy Build.* **2017**, *154*, 30–45. [CrossRef]
22. Leissner, J. The Impact of Climate Change on Historic Buildings and Cultural Property. In *UNESCO Today*; Deutsche UNESCO-Kommission: Bonn, Germany, 2011; pp. 44–45.
23. Bertolin, C.; Camuffo, D.; Leissner, J.; Antretter, F.; Winkler, M.; van Schijndel, A.W.M.; Schellen, H.L.; Kotova, L.; Mikolajewicz, U.; Brostrom, T.; et al. Results of the EU project Climate for Culture: Future climate-induced risks to historic buildings and their interiors. In Proceeding of the 2nd Annual SISC Conference, Venice, Italy, 29–30 September 2014.
24. Hagentoft, C.-E.; Kalagasidis, A.S.; Adl-Zarrabi, B.; Roels, S.; Carmeliet, J.; Hens, H.; Grunewald, J.; Funk, M.; Becker, R.; Shamir, D.; et al. Assessment method of numerical prediction models for combined heat, air and moisture transfer in building components: Benchmarks for one-dimensional cases. *J. Therm. Envel. Build. Sci.* **2004**, *27*. [CrossRef]
25. Hagentoft, C.-E. *Introduction to Building Physics*; Studentlitteratur: Lund, Sweden, 2001; ISBN 91-44-01896-7.

26. Künzel, H. Verfahren zur ein- und Zweidimensionalen Berechnung des Gekoppelten Wärme- und Feuchtetransports in Bauteilen mit Einfachen Kennwerten. Ph.D. Thesis, Universität Stuttgart, Stuttgart, Germany, July 1994. (In German)

27. Holl, K.K. Der Einfluss von Klimaschwankungen auf Kunstwerke im historischen Kontext. Untersuchung des Schadensrisikos Anhand von Restauratorischer Zustandsbewertung, Laborversuchen und Simulation. Ph.D. Thesis, Technical University, Munich, Germany, July 2016. (In German)

28. Nik, V.M. Hygrothermal Simulations of Buildings Concerning Uncertainties of the Future Climate. Ph.D. Thesis, Chalmers University of Technology, Gothenburg, Sweden, May 2012.

29. Samuelsson, P.; Gollvik, S.; Jansson, C.; Kupiainen, M.; Kourzeneva, E.; van de Berg, W.J. *The Surface Processes of the Rossby Centre Regional Atmospheric Climate Model (RCA4)*; Swedish Meteorological and Hydrological Institute (SMHI): Norrköping, Sweden, 2015.

30. Nik, V.M. *Climate Simulation of An Attic Using Future Weather Data Sets—Statistical Methods for Data Processing and Analysis*; Chalmers University of Technology: Gothenburg, Sweden, March 2010.

31. Jacob, D.; Mikolajewicz, U.; Kotova, L. *Climate for Culture—WP1: Assessment Report on Climate Evolution Scenarios Relevant for the Selected Regions*; Deliverable 1.1, Internal Project Report; Danish Meteorological Institute: Copenhagen, Denmark, 2002.

geosciences

MDPI

Article

How Can Climate Change Affect the UNESCO Cultural Heritage Sites in Panama?

Chiara Ciantelli [1], Elisa Palazzi [2], Jost von Hardenberg [2], Carmela Vaccaro [3], Francesca Tittarelli [4] and Alessandra Bonazza [1,*]

[1] Institute of Atmospheric Sciences and Climate, Italian National Research Council, via Gobetti 101, 40129 Bologna, Italy; c.ciantelli@isac.cnr.it
[2] Institute of Atmospheric Sciences and Climate, Italian National Research Council, Corso Fiume 4, 10133 Turin, Italy; e.palazzi@isac.cnr.it (E.P.); j.vonhardenberg@isac.cnr.it (J.v.H.)
[3] Department of Physics and Earth Sciences, University of Ferrara, Via Giuseppe Saragat, 1, 44124 Ferrara, Italy; vcr@unife.it
[4] Department of Materials, Environmental Sciences and Urban Planning—SIMAU, Università Politecnica delle Marche, Via Brecce Bianche 12, 60131 Ancona, Italy; f.tittarelli@univpm.it
* Correspondence: a.bonazza@isac.cnr.it; Tel.: +39-051-639-9571

Received: 13 July 2018; Accepted: 3 August 2018; Published: 7 August 2018

Abstract: This work investigates the impact of long-term climate change on heritage sites in Latin America, focusing on two important sites in the Panamanian isthmus included in the World Heritage List: the monumental site of Panamá Viejo (16th century) and the Fortresses of Portobelo and San Lorenzo (17th to 18th centuries). First of all, in order to support the conservation and valorisation of these sites, a characterisation of the main construction materials utilized in the building masonries was performed together with an analysis of the meteoclimatic conditions in their vicinity as provided by monitoring stations recording near-surface air temperature, relative humidity, and rainfall amounts. Secondly, the same climate variables were analysed in the historical and future simulations of a state-of-the-art global climate model, EC-Earth, run at high horizontal resolution, and then used with damage functions to make projections of deterioration phenomena on the Panamanian heritage sites. In particular, we performed an evaluation of the possible surface recession, biomass accumulation, and deterioration due to salt crystallisation cycles on these sites in the future (by midcentury, 2039–2068) compared to the recent past (1979–2008), considering a future scenario of high greenhouse gas emissions.

Keywords: built heritage; environmental impact; damage functions; Central America; surface recession; biomass accumulation; salt crystallisation

---

1. Introduction

Awareness of the possible risks connected with climate change and its impacts on the environment and society is increasingly growing. In recent years, particularly in Europe, the Cultural Heritage sector has become conscious of the potential problems related to climate change impacts on the materials which constitute our "tangible culture". Indeed, the Intergovernmental Panel on Climate Change (IPCC) mentioned for the first time the "Cultural Heritage" issue in its Fifth Assessment Report (AR5, 2014) [1]. In particular, Section 3.4.3 of the WG3 highlighted the need to promote people's wellbeing, including the loss of cultural heritage sites, as a metric of quality of life [1,2].

The importance of identifying the multiple effects of climate change on cultural heritage has started to be recognized even in the political sector. In fact, a recent publication commissioned by the European Commission (EC) addressed the safeguarding of cultural heritage from natural and anthropic disasters [3]. In particular, that work was developed with the aim of providing guidelines and recommendations for possible measures to improve the risk management of cultural heritage at the European level and in support of the implementation of the Sendai Framework Action Plan Priority 4 [4].

Nevertheless, further steps are necessary in order to ensure a sustainable culture, for assuring the best preservation of monumental complexes, archaeological sites, historical urban centres, artefacts, etc., and for guaranteeing their fruition by the future generations.

With the aim of enhancing knowledge in this field, in particular in Central America, an area not yet largely investigated, an international research effort was initiated in 2014 [5]. In this context, the present study considers two sites of global importance: the monumental site of Panamá Viejo, the first Spanish settlement on the Pacific Coast, and the Spanish military fortifications arising on the northern littoral of the isthmus (Figure 1). Both are included in the List of World Heritage Sites; the fortifications on the Caribbean Side of Panama, in particular, are on the List of World Heritage in Danger since 2012. Furthermore, the Portobelo and San Lorenzo Fortifications are in rural areas, while Panamá Viejo is in the urban site of Panama City; nevertheless, both sites are exposed to the same climatic conditions.

Figure 1. Pictures of different monuments present at the sites: (**a,b**) at Panamá Viejo, respectively, Convento de las Monjas de la Concepción and Torre de la Catedral; (**c**) Fort San Lorenzo; (**d**) Fort San Jeronimo, at Portobelo.

Recognising the potential contribution of climate change in increasing the vulnerability of natural, economic, and social systems, the Panamanian Government in 2007 approved the National Policy on Climate Change (Executive Decree No. 35, 26 February 2007). Seven years later, at the United Nations Climate Summit, the Panamanian President Varela Rodríguez presented the Estrategia Nacional de Cambio Climático de Panamá—ENCCP (Panama's strategy for challenging Climate Change), which developed the commitments of environmental protection measures established in the Plan Estratégico de Gobierno 2015–2019 (PEG) [6,7].

The PEG addresses the issue of adaptation to climate change impacts, considering several sectors, such as agriculture, water, energy, and logistics, marine-coastal areas, and resilient districts but neglecting cultural heritage. Thus, the present study aims to contribute to increasing the resilience of UNESCO cultural heritage sites in Panama to the environmental impacts of climate change.

1.1. Overview of Environmental Impact on Cultural Heritage

In the last decades, several works have dealt with the issue of the weathering of building materials, as described by a recent publication of Camuffo [8], which presents a comprehensive picture based on the literature of the damage occurring to heritage construction materials due to the impact of weather and air pollution, acting separately or in synergism.

Considering the stone masonry, the first step for evaluating the damages is the characterisation of the material, which includes the analysis of its composition, mineralogy, petrography, and physical features (e.g., porosity), all being necessary to understand its current state of conservation. Furthermore, we have to consider that the deterioration phenomena can be modified (accelerated/slowed, increased/decreased, etc.) by climate change [9,10].

Several damage or dose-response functions have been developed in the literature to reproduce the processes of change that a material can undergo if exposed to particular conditions. As Strlič et al. [11] note, "Damage functions can be defined as functions of unacceptable change to heritage dependent on agents of changes", thus, mathematical expressions can be considered instruments for modelling materials modifications, allowing us to evaluate past and future damages.

Regarding future damages, it is possible to implement "predictive maintenance" based on future climate projections performed with numerical models of the climate system [11], which are mathematical formulations "constructed from studies of the current climate system, including atmosphere, ocean, land surface, cryosphere and biosphere, and the factors that influence it such as greenhouse gas emissions and future socio-economic patterns of land use" [12].

Indeed, extracting from the models the parameters that act as drivers of changes on cultural heritage, such as rainfall, relative humidity, and surface air temperature, it is possible to apply them in dose-response functions, representing different kinds of material damages. In order to validate climate models before their variables are used in damage functions and equations, the simulated data should be compared with measured ones.

Several studies have been performed in the last decades based on the application of dose-response functions for estimating the damages on stone materials belonging to monuments in Europe. Dividing the equations in typologies of the measured deterioration process, we can consider damages that cause loss of material, as thermoclastism (thermal stress) affecting marbles [13], surface recession due to rainfall action on carbonate buildings [14–16], salt transitions [17], and cryoclastism (freeze–thaw effect) [18]. While, referring to accumulation of material, functions have been created for soiling and blackening phenomena [19] and biodeterioration [20]. Existing damage functions that describe climate change impacts on cultural heritage materials mainly account for the effects of slow changes, since the effect of extreme events have a higher degree of uncertainty associated.

In 2012, Sabbioni et al. [21] compiled a vulnerability atlas and guidelines for the protection of European cultural heritage (in particular, for the outdoor heritage) against climate change effects, collecting all of the results produced by the FP6 Noah's Ark Project (2004–2007). The climate parameters, utilized for mapping future scenarios, were extracted from two climate models developed at the Hadley Center (UK), i.e., the global Coupled Climate Model (HadCM3) and the Regional Climate Model (HadRM3) with spatial resolutions of, respectively, 295 km × 298 km and 50 km × 50 km. This allowed the authors to produce maps describing climate conditions, damage, and risk for cultural heritage at the European level. The FP6 Noah's Ark Project was then the basis for another project, namely, the FP7 Climate for Culture Project—CfC (2009–2014), which addressed forecast hazard and damage projections to assess the impact of slow, ongoing climate change on historic buildings. In that case, a regional hydrostatic climate model was used, called REMO, which has a spatial resolution of approximately 10 km × 10 km for the European and Mediterranean areas. The CfC FP7 project produced more than 55,000 maps to assess the vulnerability of built heritage as well as of artworks preserved indoors [22–27].

Within this research, we analysed a set of simulations performed with the EC-Earth global climate model run at a particularly high spatial resolution (25 km × 25 km, more details in Section 2.1), to map deterioration phenomena due to slow climate change for both past and future conditions over the Panamanian isthmus area encompassing the different locations studied in this work.

1.2. Panamanian Climate Classification

According to the Köppen–Geiger climate classification on the "thermal zones of the Earth" [28,29], then validated for the second half of the 20th century by Kottek et al. [30], the area of Panama has been classified as an equatorial climate, within the types listed below:

- Af: "Equatorial rainforest, fully humid";
- Am: "Equatorial monsoon";
- Partially, Aw: "Equatorial savannah with dry winter";

In this classification, the first letter indicates the main climate typology, in our case A, which refers to the equatorial zone showing a minimum temperature greater than or equal to +18 °C. The second letter indicates precipitation conditions; in our case, "f" refers to the criterion of the precipitation of the driest month (P_{min}) $\geq$ 60 mm/month, while "m" refers to the criterion of the accumulated annual precipitation (P_{ann}) $\geq$ 25 (100−P_{min}) mm/year. Finally, "w" refers to the criterion of $P_{min} <$ 60 mm/month in winter [30].

In 2000, a Panamanian geographer and historian, Dr Alberto McKay [31], developed a specific classification of the isthmic environment and climate which takes into account the influence of the oceanic masses and has led to the identification of seven different climate typologies for Panama. Two of them can be associated with the sites that are studied in the present work. The Tropical Oceanic Climate with a short dry season characterises Portobelo and San Lorenzo, where average temperature values are around 25.5 °C on the coastal area and 26.5 °C in the continental part. Precipitations are abundant, reaching annual amounts of 4760 mm. The dry season has a brief duration of 4–10 weeks, with 40–90 mm of precipitations between February and March. While Panamá Viejo has a Tropical Climate with a prolonged dry season, featuring the warmest climate with average temperature values of 27–28 °C and recording the lowest amount of precipitation, lower than 2500 and 1122 mm in a few areas. The long-lasting dry season (approximately from December to March) is characterised by strong winds, with mid-high clouds, low humidity, and, consequently, high evaporation.

The proximity of our study sites to the sea made sea level and tidal variations important factors for the progressive coastal erosion in the last five centuries [32–34]. This carried the ruins of Panamá Viejo to be only 100 m from the sea [35], causing also the erosion of some monuments at the site of Panamá Viejo erected in the part closest to the coast. Furthermore, tidal variations expose the fortresses located in Portobelo to seawater action, in addition to the rainfall effects. Indeed, several of these structures are exposed to sea intrusion, especially during storms, while the daily tidal variations cause the erosion of the external walls and water infiltration. Even if the short-term tidal effects and coastal erosion are not the subject of this study, their long-term consequences can contribute to the presence of salts within the masonries and thus to their weathering action [36]. In fact, by the end of the 21st century, it is very likely that a large fraction of the coastal areas around the globe will be affected by sea level rise induced by global warming [37], with specific impacts strongly varying from coast to coast as a result of a combination of different factors.

2. Materials and Methods

2.1. Selection of Monitoring Stations and Climate Model

With the purpose of obtaining measured climate parameters in the surroundings of the monumental site of Panamá Viejo as well as the Portobelo and San Lorenzo areas in the recent past, several monitoring stations were selected. Specifically, they belong to the networks of the Panama Canal Authority (ACP) and of the Empresa de Trasmision Electrica S.A. (ETESA), and their data were downloaded from the Smithsonian Tropical Research Institute (STRI) and ETESA websites, respectively [38,39]. It has to be underlined that the data presented here, elaborated by authors, are a modified version of the original. The name, elevation, and measured parameters (rainfall, R; relative humidity, RH; surface air temperature, T) of the various stations considered in this study are listed in Table 1 and their geographical locations are shown in Figure 2. As suggested by STRI, the Cristobal, Cocosolo, and Limon Bay stations (CCL in the legend of Figure 2) are grouped together and considered as a single station.

Table 1. Monitoring stations selected as the closest to the sites of interest, respectively: in white, Panamá Viejo; in blue, Portobelo; and in green, San Lorenzo.

Name	Label	Elevation (m)	Date	Available Climate Parameters		
				R	RH	T
Hato Pintado	HP	45	1 July 1987–	×		
Tocumen	T	18	1 January 1970–1 January 2013	×	×	×
Balboa FAA	B-FAA	10	1 January 1908–	×	×	×
Portobelo	Pb	2	1 May 1908–1 January 2004	×		×
Gatun Rain Z.C.	G	31	1 January 1905–	×	×	×
Cristobal		8.5	1 October 1862–30 September 1979			
Coco Solo	CCL	4.6	1 September 1980–30 June 1996	×	×	×
Limon Bay		3	1 November 1996–	×	×	×
Fort Sherman	FS	9	24 April 1997–14 October 2014	×	×	×
Pina	P	3	1 December 1970–1 November 1998	×		

In this study, specific simulations with the state-of-the-art global climate model EC-Earth [40,41] performed in the framework of the "Climate SPHINX" (Stochastic Physics HIgh resolutioN eXperi-ments) experiment were also considered [42,43]. EC-Earth is a Earth system model developed by a consortium of European research institutions and universities, based on state-of-the-art models for the atmosphere, ocean, sea ice, and the biosphere [42]. The simulations performed with EC-Earth in Climate SPHINX were atmosphere-only experiments extending 30 years into the past (from 1979 to 2008) and 30 years into the future (from 2039 to 2068) using forcing conditions from the Representative Concentration Pathway emission scenario RCP 8.5 [44]. EC-Earth was run exploring five different horizontal resolutions, i.e., ~125, ~80, ~40, ~25, and ~16 km. For the present study, only the simulations performed at 25 km were analysed.

The area of the central part of the isthmus covered by the EC-Earth pixels is shown in Figure 2. The following model variables—daily rainfall (mm), daily surface air temperature (°C), and daily relative humidity (RH)—were analysed only in the model pixels where the monitoring stations are located.

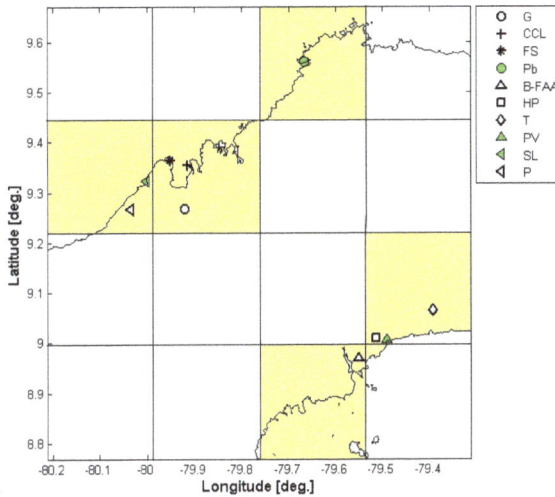

Figure 2. Grid areas of EC-Earth model. Highlighted in yellow are the zones that overlap or are near the sites of interest. On the right, for the legend of the markers, the white symbols indicate the monitoring stations, the green markers the sites, respectively: G = Gatun Rain Z.C., CCL = Cristobal–Cocosolo–Limonbay, FS = Fort Sherman, Pb = Portobelo (both monitoring station and site), B-FAA = Balboa-FAA, HP = Hato Pintado, T = Tocumen, PV = Panamá Viejo, SL = San Lorenzo, P = Pina.

2.2. Analysis of Construction Materials

Specimens collected from the masonries and possible quarries during a sampling campaign underwent the following mineralogical–petrographic, physical, and chemical analyses:

- *Stereomicroscope observations* using an Optika SZ6745TR equipped with a webcam, MOTICAM 2005 5.0 Mp, and Moticam Image Plus 2.0 software, were utilized for performing preliminary analyses of the bulk samples.
- *Polarized light microscopy (PLM) investigations* were performed for the analysis of uncovered thin sections, partially polished in order to observe in both transmitted and reflected light, using an Olympus BX 51 microscope equipped with scanner-camera and the MICROMAX software "Primoplus_32" vers. 8.11.02. Furthermore, for evaluating the state of conservation, the thin sections were realized through transversal cuts from the external to the inner part of the samples.
- *X-ray powder diffraction (XRPD) analysis* was performed for determining the mineralogical phases present using a Philips PW 1730 diffractometer equipped with a copper anticathode and a nickel filter. The measurement conditions have a diffraction interval of 2θ, between $5°$ and $50°$, and a $2°/\text{min}$ step at 40 kV voltage and 30 mA current intensity. In addition, further analyses were performed in order to verify the clay minerals present in several samples, utilizing a Bruker AXS D8, in Bragg-Brentano geometry, equipped with an X-ray tube and a SolX solid state detector, working in low-temperature through Peltier cooling system. The samples for these techniques underwent a powdering process, utilizing two mills, firstly a jaw crusher and secondly a mortar grinder with agate jar and pestle. For materials showing high hardness, the process was finished by manual grinding with an agate mortar.
- *Environmental scanning electron microscopy and microchemical investigations (ESEM-EDS);* analyses were carried out to determine the elemental composition of specific areas of interest, already observed by stereomicroscope and PLM investigations on both bulk and thin sections. The instrument utilized a ZEISS EVO LS 10 with LaB6 source.

- *X-ray fluorescence (XRF)* was performed on powder pellets (see XRPD section for powdering process) pressed with boric acid powder as a binder using a wavelength-dispersive automated ARL Advant'X spectrometer. Accuracy and precision for major elements were estimated 2%–5%; trace elements (above 10 ppm) were estimated at greater than 10%.
- *Mercury intrusion porosimetry (MIP)* was performed to understand the porosimetry features of the materials and to index of their state of conservation. Specimens that showed enough material (~1 cm × 1 cm × 1 cm to 2 cm × 2 cm × 2 cm) were selected and analysed by a porosimeter, PASCAL 240, THERMO SCIENTIFIC.
- *Ion chromatography (IC)* analyses were performed on samples showing particular patina or superficial alteration phenomena in order to evaluate the possible presence of soluble salts present in the masonry. The selected specimens, powdered (see XRPD section for powdering process) and then solubilized (sample/water ratio 25 mg/~50 g), were investigated by a DIONEX ICS 900. Anions analysis: Column S23 Pre-column G23; cations analysis: Column CS12 Pre-column CG12.

3. Results and Discussion

3.1. Materials Characterisation

Optical microscopy integrated with X-ray diffraction and X-ray fluorescence showed that construction materials at Panamá Viejo masonries are mainly composed of polygenic breccias, tuffites, basaltic andesites, rhyolites, and sporadic rhyodacites. For the Portobelo Fortifications, coral limestones and sandstones were identified as the principal construction materials, while basaltic andesite was observed only at Fort San Fernando [45]. Finally, at Fort San Lorenzo, tuffites and grainstones were detected in the masonries (Figure 3).

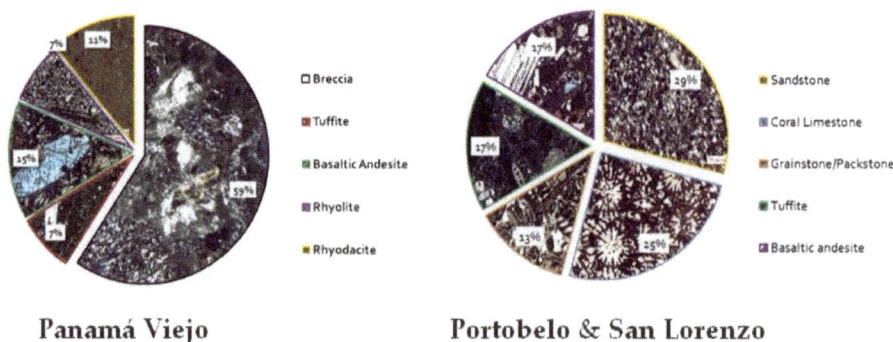

Panamá Viejo   **Portobelo & San Lorenzo**

Figure 3. Lithotypes detected in the sites of Panamá Viejo (on the **left**) and of Portobelo and San Lorenzo (on the **right**). The pie charts indicate the percentage of materials detected with respect to the entire number of samples analysed at each site. Each section of the pie charts shows a picture obtained by polarized light microscopy (PLM) observations of thin sections of the different lithotypes. The percentage is obtained considering as 100% the total number of samples per site.

Concerning the state of conservation, the most diffused deterioration phenomena are due to biological growth, material loss, disintegration (sanding and pulverization), salt encrustations (especially at the Portobelo Fortifications, presenting a process of calcite crystallisation developed in several steps), the presence of soluble salts, and chromatic alteration (in particular, affecting rhyolites) (Figure 4).

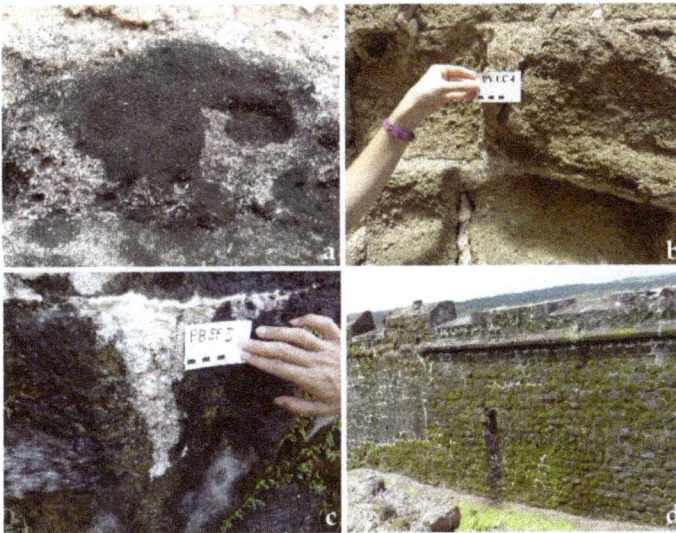

Figure 4. Pictures representing several deterioration processes observed at the sites. (**a,b**) at Panamá Viejo, (**a**) biological growth; (**b**) surface recession/material loss; (**c**) encrustation at Fort San Fernando (Portobelo); (**d**) biological growth at Fort San Lorenzo.

Through the petrographic characterisation performed by PLM, porosity was also investigated. In particular, coral boundstones and grainstones/packstones show a very high intergranular and intragranular porosity. While, considering the sandstones and tuffites, they are more compact but affected by cracks. MIP analysis revealed, in general, that the majority of samples analysed present an average pore diameter lower than the 0.2 μm, the threshold of micropores. Furthermore, almost all samples show a range of pore distribution between 0.01 and 10 μm, except for rhyodacites and rhyolites, which tend to have the majority of pore diameters between 0.01 and 1 μm and coral limestones, showing a predominance towards to 10 μm. Regarding in particular carbonate-based stones, the percentage of accessible porosity percentage under 25% was detected in several samples of polygenic breccia, tuffite, sandstone, grainstone, and boundstone.

Regarding the presence of soluble salts, through the IC investigations, it was possible to define calcium as the most abundant cation in all specimens analysed and at every site (Table 2). Moreover, high presence of sulphates and nitrates has been detected, particularly at Panamá Viejo, since it is within an urban area in proximity to a high-traffic road and thus affected by heavy anthropogenic pollution. At Portobelo and San Lorenzo, the most abundant anion is chloride, and, considering the cations detected, it can form sodium, potassium, ammonium, and magnesium chloride. Nevertheless, the presence of halite is assumed at every site since chloride is largely present also in Panamá Viejo specimens. Furthermore, in presence of water, Cl^- can create hydrochloric acid, as sulphates and nitrates can form, respectively, sulphuric and nitric acid, causing the dissolution of carbonates, which can recrystallise inside the stone porosity and cause internal tensions or on the surface, forming superficial encrustations of calcite (Figure 4c).

For detailed information regarding the previous listed analyses (Section 2.2) and related results, see Ciantelli, 2017 [5].

Table 2. Ions detected by ion chromatography (IC) in representative lithotypes belonging to Panamá Viejo, Portobelo, and San Lorenzo, (Pol. Br. = polygenic breccia; Tuff. = tuffite; Rhyod = rhyodacite; Bas. And. = basaltic andesite; Sandst. = sandstone; Grainst. = grainstone; Cor. Lim. = coral limestone). In the columns, we highlighted in grey the most abundant anions and cations.

							Panamá Viejo							
			Anions Concentration/ppm							Cations Concentration/ppm				
Lithot.		$C_2H_3O_2^-$	PO_4^{3-}	$C_2O_4^=$	CHO_2^-	NO_2^-	NO_3^-	$SO_4^=$	Cl^-	NH^{4+}	K^+	Mg^{2+}	Na^+	Ca^{2+}
	Min	3	0	0	25	0	39	162	382	0	0	15	287	0
Pol. Br.	Mean	18	6	45	40	5	1034	2140	1316	33	212	458	2414	3756
	Max	87	42	319	68	11	3551	11,525	3821	89	354	1123	10,076	27,884
Tuff.		0	0	34	75	62	6352	3098	10,675	99	237	294	5306	33,039
	Min	4	143	0	17	26	1033	634	1334	25	244	288	1518	1775
Rhyod.	Mean	546	171	0	29	27	1042	705	1537	30	247	321	2595	1781
	Max	1088	199	0	41	27	1051	775	1739	35	250	353	3672	1787
							Portobelo & San Lorenzo							
			Anions Concentration/ppm							Cations Concentration/ppm				
Lithot.		$C_2H_3O_2^-$	PO_4^{3-}	$C_2O_4^=$	CHO_2^-	NO_2^-	NO_3^-	$SO_4^=$	Cl^-	NH^{4+}	K^+	Mg^{2+}	Na^+	Ca^{2+}
	Min	0	0	0	32	0	19	74	617	23	313	318	789	0
Bas. And.	Mean	0	0	11	57	10	40	115	664	52	453	442	1013	2017
	Max	0	0	33	103	29	61	169	704	103	641	633	1377	2888
	Min	0	0	0	49	12	58	203	729	53	219	496	453	0
Tuff.	Mean	4	0	0	52	22	165	260	1545	238	368	562	714	32,204
	Max	12	0	0	56	35	343	326	2802	498	618	604	1217	33,233
	Min	0	0	0	15	0	35	132	316	85	306	938	277	32,356
Sandst.	Mean	6	40	0	35	0	160	182	974	94	636	1066	1337	38,644
	Max	16	67	0	50	0	369	222	1779	112	1160	1208	2576	49,087
Grainst.		15	0	20	22	43	21	362	700	52	158	652	452	48,671
	Min	0	0	0	38	13	62	408	495	0	0	0	0	34,328
Cor. Lim.	Mean	10	0	8	49	35	68	529	590	159	153	147	545	46,709
	Max	28	0	24	60	66	71	634	766	255	311	231	846	59,090

3.2. Comparison between Climate Simulations and Monitoring Stations

Data obtained from the monitoring stations and from the EC-Earth model simulations have been compared over a common period of time, either the model "historical period"—30 years extending from 1979 to 2008—or shorter time periods, depending on the temporal availability of the measured data.

Figure 5a,b show, respectively, the comparison between the annual cycle climatology (averaged over the 1979–2008 time period) of surface air temperature and relative humidity measured at the Tocumen site and simulated by EC-Earth. The seasonality of both variables is well captured by the model in spite of its underestimation of surface air temperature and overestimation of relative humidity with respect to the measurement data. Figure 5c,d show a comparison between the model and observations of the rainfall climatology. At Tocumen and Hato Pintado, the model exhibits an overestimation of rainfall amounts during the wet season, in particular between May/June to September/October (Figure 5c), while on the on the North Coast, in the areas of San Lorenzo and Portobelo, it significantly underestimates rainfall amounts compared to the observations between October and December (Figure 5d).

Based on the comparison between the model outputs and the observations over the study sites, bias-correction methods have been applied to the EC-Earth outputs before using them in the damage functions. Basic bias-correction methods include an adjustment of the mean value by adding a temporally constant offset or by applying a correction factor to the simulated data. This additive or multiplicative constant quantifies the average deviation of the simulated data from the observed one over the historical period over which data was compared. In this case, additive adjustments have been adopted for bias correcting the modelled temperature, while multiplicative adjustments were used for relative humidity and rainfall, as better explained below.

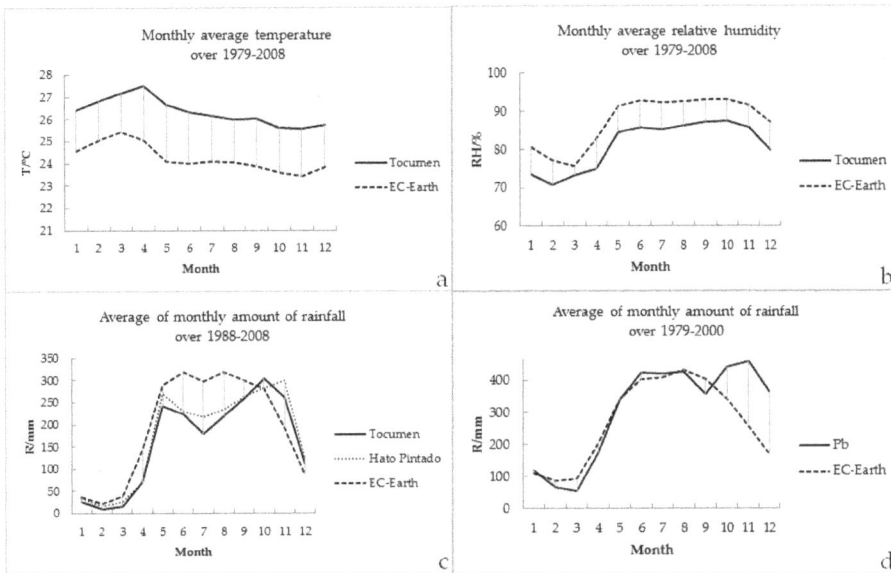

Figure 5. (**a,b**) Comparison between the monthly average values of, respectively, T and RH, over the period 1979–2008, in the area of Panamá Viejo between the data collected from the Tocumen monitoring station and those simulated by EC-Earth; (**c,d**) comparison of the average of monthly amount of rainfall: (c) in the period 1988–2008, between data of the Hato Pintado and Tocumen stations and of the EC-Earth model; (**d**) in the period 1979–2000, between data of the Portobelo (Pb) station and of EC-Earth model.

- Rainfall: Being *Po* and *Pm* the observed and modelled daily precipitation and being their long term average over the period 1979–2008 indicated by an overbar, the multiplicative correction factor, *fP*, is calculated as Equation (1):

$$fP = \frac{\overline{Po}}{\overline{Pm}}$$
(1)

Therefore, the factor *fP* is multiplied by the daily precipitation simulated by the model to obtain the model bias-corrected rainfall.

- Relative Humidity: The same procedure as for rainfall was applied to relative humidity data, using a multiplicative correction factor to bias correct the model data based on observations. In this case, the correction factor multiplied to the daily model data is *fRH* (Equation (2)):

$$fRH = \frac{\overline{RHo}}{\overline{RHm}}$$
(2)

where $\overline{RHo}$ is the average over the period 1979–2008 of the daily RH for observations and $\overline{RHm}$ is the average of the RH for the model.

- Temperature: In this case, an additive correction factor *fT* is calculated as the difference between the climatological average of the observed daily temperature over the period 1979–2008 and the climatological average of the modelled daily temperature over the same time period, as follows (Equation (3)):

$$fT = \overline{Tm} - \overline{To}$$
(3)

The bias-corrected temperature is calculated by adding the correction factor to the daily time series simulated by the model.

Correction factors for all three variables have been applied to both historical data (1979–2008) and to the future data available for the period 2039–2068.

3.3. Selection of Damage Functions

Taking into account the Panamanian climate conditions, the composition of the materials belonging to the buildings under study, their potential deterioration phenomena, and the damage functions developed in the field of stone materials conservation, the following three equations have been applied, considering slow change effects and no extreme events.

(1) Surface recession, according to Lipfert [46], is due to the effect of rain washout. In particular, for the present study, the Lipfert modified equation has been chosen, according to [15], as follows (4):

$$L = K_{1,2} \times R \tag{4}$$

where:

L: surface recession per year (μm year^{-1})
K_1: 18.8 intercept term based on the solubility of $CaCO_3$ in equilibrium with 330 ppm CO_2 (μm m^{-1})
K_2: 21.8 intercept term based on the solubility of $CaCO_3$ in equilibrium with 750 ppm CO_2 (μm m^{-1})
R: precipitation (m year^{-1})

Bonazza et al. [15] demonstrated that the clean rain effect (karst effect) was the dominant term, accounting for 50%–90% of stone loss. It has to be underlined that the K_1 constant was utilized both for the past and future situation, while the K_2 constant was applied only for the future projections, since the concentration of carbon dioxide is expected to be higher than the past of 330 ppm.

Finally, it has to be underlined that this function has been demonstrated to be valid for carbonate rocks having a porosity lower than 25%.

(2) Salt transitions cycles have been considered for halite. Indeed, considering the proximity to the sea of all the sites under study and the equilibrium RH (%) and T (°C) of NaCl corresponding to 75.3% at 25 °C and 75.1% at 30 °C, dissolution–crystallisation transformations can occur. Specifically, assuming T as a constant and according to Grossi et al. [17], the frequency of cycles has been calculated counting the number of times the average daily RH crossed the DRH/CRH of 75.3% or 75.1% on consecutive days. Only the transitions that occurred when the humidity was decreasing, therefore passing from liquid to solid state, were counted. Thus, it can be affirmed that the number of transitions is virtually the number of dissolution–crystallisation cycles.

(3) Biomass accumulation, considered as organic carbon accretion on the surfaces, it has been obtained utilizing the Gomez-Bolea et al. function (5) [20]:

$$B = \exp^{(-0.964+0.003P-0.01T)} \tag{5}$$

where the quantity of biomass B (mg) on surface unity (cm^{-2}) is obtained by applying the annual amount of precipitation P (mm) and the annual mean of temperature T (°C).

Nevertheless, it has to be mentioned that this function has been validated for horizontal surfaces of hard acid stones in nonurban European environments.

3.4. Damage Evaluation and Future Predictions

Generally, rainfall events in Panama are very intense, with quite short duration, and are prolonged during the rainy season, which covers the majority of the year (8–10 months). Recent studies demonstrated that in the last years, precipitation around Panama City has increased, exceeding long-term averages. Nevertheless, the beginning and the duration of the rainy, and, consequently, of the dry season have been changing [47]. The concentration of precipitations in specific periods (as mid-August), always according to Paton [47], causes an increment in rain intensity, triggering effects on the monuments exposed to it, such as a higher mechanical erosion and chemical deterioration due to water permeation in the structures [48].

In addition, this phenomenon, in conjunction with the growth of urbanization, which is not adequately planned, caused an increase of surface runoff. Actually, the fast proliferation of urban areas, especially overbuilding and roads paving, creates waterproof surfaces that prevent rainfall water from being drained by the ground, triggering extreme effects such as erosion, sediments, and floods. Finally, runoff can carry contaminants dissolved in it, thus representing a source of danger both for the human beings and for the cultural heritage, which can be exposed to salts and/or pollutants dissolved in water and, consequently, to salt-cycle weathering [31,49–51].

Strictly connected with these phenomena, the most diffused deterioration morphologies observed in the investigated sites are loss of materials, salts crystallisation, and biological growth. Therefore, damage functions have been selected and applied in order to evaluate and predict these alteration patterns.

Considering the surface recession, the modified Lipfert function (related to the karst effect) was utilized, as described above. In order to understand the future trend, the function has been applied both in the most optimistic (using K_1) and in the most pessimistic (using K_2) form. The EC-Earth projections by mid-21st century show an increase with respect to the past from 6–7 μm year^{-1} (calculated at Panamá Viejo) to 10–12 μm year^{-1} (at San Lorenzo and Portobelo). It can be assumed that, in the future, the sites situated at the northern Panamanian coast might be more affected by surface recession than those located at the southern coast (Figure 6).

In consideration of the salt cycles of dissolution and crystallisation, halite has been selected as a priority phase of investigation, since sodium and chloride were the most abundant ions in the stone samples at all sites because of their proximity to the sea. In addition, the thermohygrometric conditions at the sites are favourable to dissolution and crystallization cycles of halite. Therefore, the past and future monthly transitions of NaCl have been calculated, considering as an event the passage from RH values higher than 75.3% to values smaller than this threshold, counted only if it happened on consecutive days. Nevertheless, we also have to bear in mind the dependence of this process on temperature. Indeed, even if it can be considered a constant, the temperature of existence of NaCl transition at 75.3% is 25 °C. As stated by Satterthwaite [52], most of the Latin American and Caribbean cities can be affected by an increase of heat waves in the future and, in general, higher temperatures are expected. Thus, NaCl transition at 30 °C was also considered, corresponding to a 75.1% RH threshold.

Past cycles of dissolution and crystallisation of halite highlight that the higher frequency of this phenomenon is recorded during the dry season (end of November/December to April/beginning of May). However, since we noticed an underestimation of EC-Earth simulation during the rainy season in comparison with the monitoring station data, sporadic events cannot be excluded during this period. Nevertheless, we have to bear in mind that the monitoring station represents a punctual situation, while the model grid point can be interpreted as being representative for an area.

In general, making a comparison between the past and future conditions, the period of cycles tends to maintain the seasonality trend presented in the past situation, having a slight decrease of the maximum values of cycles during the dry months and a slight increase during the rainy ones (±1). Evaluating the difference among the sites, the area near San Lorenzo seems to be the most affected one and especially during the rainy season, reaching 3–4 cycles per month, in contrast with the other sites (e.g., Figure 7).

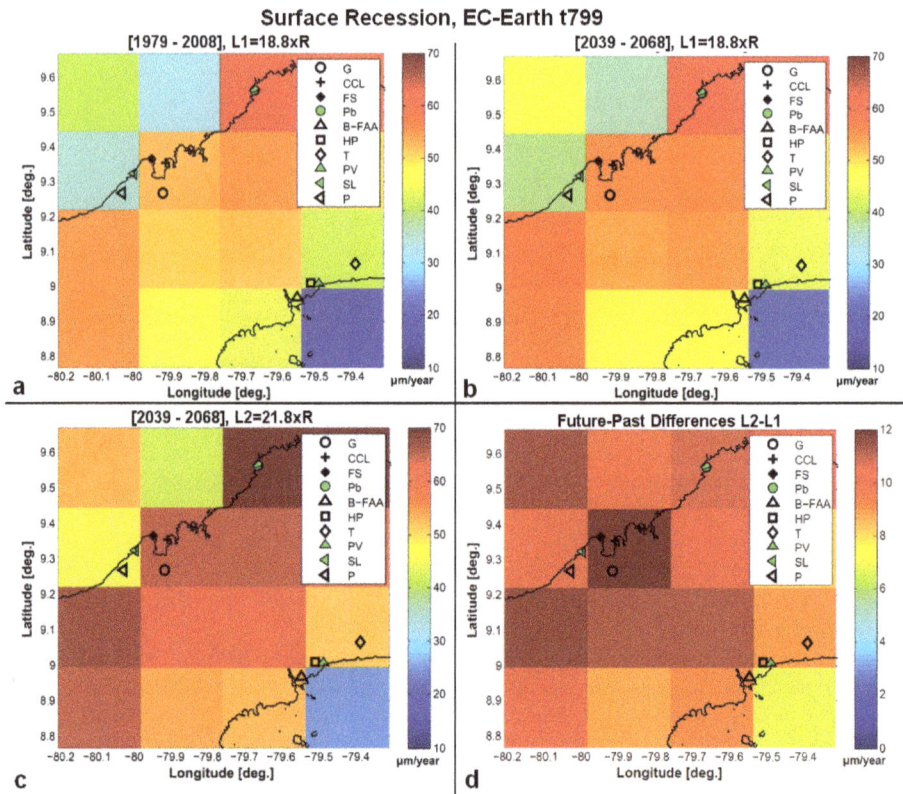

Figure 6. Estimated surface recession under EC-Earth scenario utilizing data without the bias correction. (**a,b**) (L1 = 18.8 × R): (**a**) For the past period 1979–2008 (baseline); (**b**) For the middle future period (2039–2068); (**c,d**) Middle future period (2039–2068): (**c**) Estimated surface recession (L2 = 21.8 × R) under EC-Earth scenario, (**d**) Differences between the estimated future surface recessions under EC-Earth scenario K_2 constant and past recession applying the K_1 constant: (L2future = 21.8 × R)–(L1past = 18.8 × R).

In order to estimate the biomass accumulation on hard acid stones, calculated considering the organic carbon accretion on the surfaces, the function developed by Gomez-Bolea et al. [20] has been applied. Regarding the past trend of biomass accumulation calculated with model data and compared with the biomass estimated with monitoring station records, EC-Earth represents better the range of values of the Panamá Viejo zone and the area near San Lorenzo.

In the future situation, the estimation of biomass shows at the Portobelo area the highest values (reaching 21 g cm^{-2}), followed by the area near San Lorenzo (up to 10–20 g cm^{-2} by EC-Earth simulation data), then the area near Panamá Viejo (up to ~12 g cm^{-2}), and finally the Panamá Viejo and San Lorenzo zones (1–5 g cm^{-2}). Therefore, the highest differences between future and past are recorded at the areas on the north shore, especially Portobelo (Figure 8).

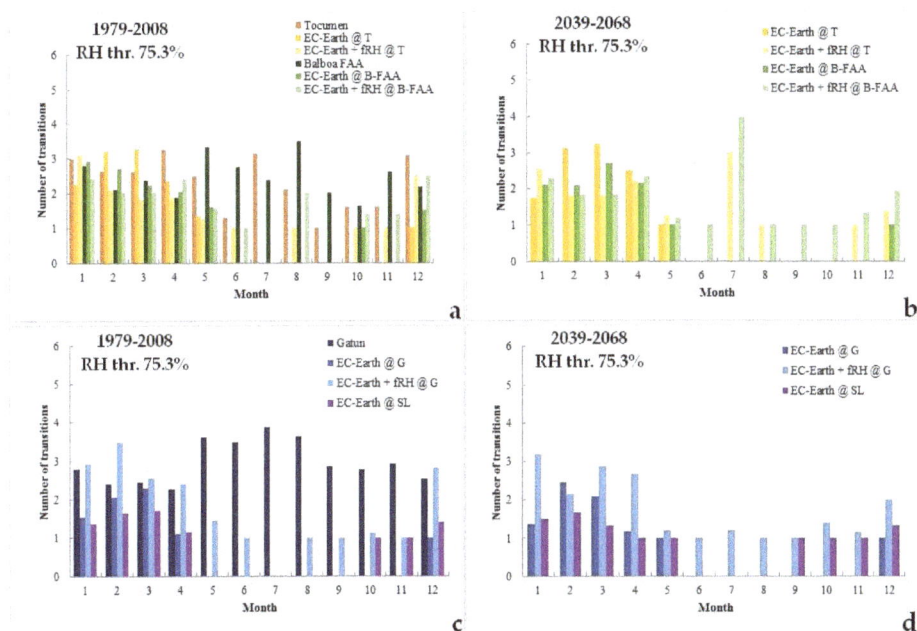

Figure 7. (**a**,**b**) Areas surrounding Panamá Viejo; (**c**,**d**) areas surrounding San Lorenzo. Comparison of mean of past (**a**,**c**) and future (**b**,**d**) NaCl monthly transitions considering 75.3% as RH threshold.

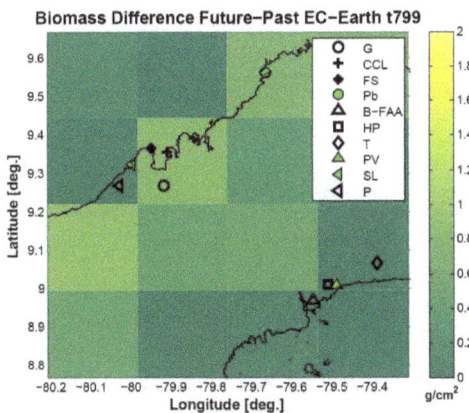

Figure 8. Difference between the estimated biomass accumulation between the future and the past period under the EC-Earth scenario, utilizing data without the bias correction.

4. Conclusions and Future Perspectives

Considering the construction materials present in the built heritage which has been investigated in the present study and the climate parameters involved in the deterioration processes, the results discussed in the previous sections allowed us to predict the possible impacts of future climate change on the studied heritage sites in Panama.

We focused on the analysis of rainfall, relative humidity, and surface air temperature, the changes of which are key drivers of the deterioration of cultural heritage. We applied future model predictions of these variables in functions to study the different kinds of material damages which might occur in the future. In particular, all functions that we considered indicate an increase of surface recession, biomass accumulation, and cycles of dissolution and crystallisation of halite in the future (2039–2068) with respect the past (1979–2008), especially in the North Coast, as shown by the analysis performed at the San Lorenzo and Portobelo areas. Nevertheless, the Panamá Viejo zone also shows an increment of surface recession and biomass accumulation, while, considering the salt cycles, growth is projected to be reduced.

This work represents an important contribution to better understand the possible future impact of climate change on the heritage sites of Central America and to support their management, restoration, and preservation. Nevertheless, in order to carry out a deeper study of the environmental impact, it will also be necessary to integrate the monitoring with pollution data, organizing in situ measurements of both climate and pollution parameters. This is fundamental also for planning field-exposure tests to strengthen the validity of the damage functions in situ. Moreover, these kinds of experiments are necessary as well for the possible evaluation of the efficiency of maintenance and restoration materials applied on specimens of the different rocks present at the sites exposed to the same conditions as the monuments.

Furthermore, since biological growth was one of the most diffused damage phenomena, a study focusing on the identification of the different kinds of species colonising the material surfaces should be implemented, with the additional purpose of improving the effectiveness of the biomass accumulation equation.

Finally, it would also be important to analyse future projections of regional sea level in the Panamanian area to understand the extent to which this could combine with other drivers and contribute to the damage of the coastal built heritage in this area.

Author Contributions: Conceptualization, A.B.; Formal analysis, C.C.; Funding acquisition, C.V. and A.B.; Investigation, C.C.; Project administration, A.B.; Resources, E.P., J.v.H., C.V., F.T. and A.B.; Supervision, A.B.; Writing—original draft, C.C.; Writing—review & editing, E.P., J.v.H. and A.B.

Funding: This research received no external funding.

Acknowledgments: This study was carried out thanks to a fruitful PhD collaborative project that involved the Institute of Atmospheric Sciences and Climate, ISAC-CNR (Bologna, Italy), the Department of Physics and Earth Sciences of the University of Ferrara (Italy) and two Panamanian Patronages: the Patronato Panamá Viejo and the Patronato de Portobelo y San Lorenzo. In particular, authors are very grateful to all the personnel of the Patronages, with a special mention to Julieta de Arango and Silvia Arroyo, belonging to the Patronato Panamá Viejo. While at the Patronato de Portobelo y San Lorenzo we want to acknowledge principally Rodolfo A. Suñé and Wilhelm Franqueza. Finally, a particular acknowledgment to the Smithsonian Tropical Research Institute (STRI's) Physical Monitoring Program, and specifically to Steven Paton, and to the Empresa de Transmisión Eléctrica, S.A.—ETESA, for allowing authors the access to the historical climate Panamanian data.

Conflicts of Interest: The authors declare no conflict of interest.

References

1. Intergovernmental Panel on Climate Change (IPCC). *Climate Change 2014: Mitigation of Climate Change*; Contribution of Working Group III to the Fifth Assessment. Report of the Intergovernmental Panel on Climate Change; Edenhofer, O., Pichs-Madruga, R., Sokona, Y., Farahani, E., Kadner, S., Seyboth, K., Adler, A., Baum, I., Brunner, S., Eickemeier, P., et al., Eds.; Cambridge University Press: Cambridge, UK; New York, NY, USA, 2014; p. 1454, ISBN 978-1-107-05821-7.

2. Schneider, S.H.; Kuntz-Duriseti, K.; Azar, C. Costing non-linearities, surprises, and irreversible events. *Pac. Asian J. Energy* **2000**, *10*, 81–106.

3. Bonazza, A.; Maxwell, I.; Drdácky, M.; Vintzileou, E.; Hanus, C.; Ciantelli, C.; De Nuntiis, P.; Oikonomopoulou, E.; Nikolopoulou, V.; Pospíšil, S.; et al. *Safeguarding Cultural Heritage from Natural and Man-Made Disasters A Comparative Analysis of Risk Management in the EU*; European Union: Brussels, Belgium, 2018; ISBN 978-92-79-73945-3.

4. European Commission. *Commission Staff Working Document—Action Plan on the Sendai Framework for Disaster Risk Reduction 2015–2030, A Disaster Risk-Informed Approach for All EU Policies*; European Commission: Brussels, Belgium, 2016.

5. Ciantelli, C. *Environmental Impact on UNESCO Heritage Sites in Panama*; University of Ferrara: Ferrara, Italy, 2017.

6. Consejo de la Concertación Nacional para el Desarrollo. *Plan Estratégico Nacional con Visión de Estado Panamá 2030*; Consejo de la Concertación Nacional para el Desarrollo: Panama, Panama, 2017; ISBN 978-9962-663-33-1.

7. Ministerio de Ambiente. Gobierno de la Republica de Panamá Estrategia Nacional de Cambio Climático de Panamá. Available online: http://www.miambiente.gob.pa/images/stories/documentos_CC/Esp_Info_V.1_ENCCP_15.12.2015.pdf (accessed on 25 June 2018).

8. Camuffo, D. Weathering of Building Materials. In *Urban Pollution and Changes to Materials and Building Surface*; Brimblecombe, P., Ed.; Imperial College Press: London, UK, 2016; pp. 19–64.

9. Abd El-Aal, A.K. Climate Change and its Impact on Monumental and Historical Buildings towards Conservation and Documentation Ammon temple, Siwa Oasis, Egypt. *J. Earth Sci. Clim. Chang.* **2016**, *7*, 339. [CrossRef]

10. Brimblecombe, P. Refining climate change threats to heritage. *J. Inst. Conserv.* **2014**, *37*, 85–93. [CrossRef]

11. Strlič, M.; Thickett, D.; Taylor, J.; Cassar, M. Damage functions in heritage science. *Stud. Conserv.* **2013**, *58*, 80–87. [CrossRef]

12. Cassar, M.; Pender, R. The impact of climate change on cultural heritage: Evidence and response. In *ICOM Committee for Conservation: 14th Triennial Meeting*; Verger, I., Ed.; James & James: The Hague, The Netherlands, 2005; pp. 610–616.

13. Bonazza, A.; Sabbioni, C.; Messina, P.; Guaraldi, C.; De Nuntiis, P. Climate change impact: Mapping thermal stress on Carrara marble in Europe. *Sci. Total Environ.* **2009**, *407*, 4506–4512. [CrossRef]

14. Kucera, V.; Tidblad, J.; Kreislova, K.; Knotkova, D.; Faller, M.; Reiss, D.; Snethlage, R.; Yates, T.; Henriksen, J.; Schreiner, M.; et al. UN/ECE ICP Materials Dose-response Functions for the Multi-pollutant Situation. *Water Air Soil Pollut. Focus* **2007**, *7*, 249–258. [CrossRef]

15. Bonazza, A.; Messina, P.; Sabbioni, C.; Grossi, C.M.; Brimblecombe, P. Mapping the impact of climate change on surface recession of carbonate buildings in Europe. *Sci. Total Environ.* **2009**, *407*, 2039–2050. [CrossRef] [PubMed]

16. Inkpen, R.; Viles, H.; Moses, C.; Baily, B. Modelling the impact of changing atmospheric pollution levels on limestone erosion rates in central London, 1980–2010. *Atmos. Environ.* **2012**, *61*, 476–481. [CrossRef]

17. Grossi, C.M.; Brimblecombe, P.; Menéndez, B.; Benavente, D.; Harris, I.; Déqué, M. Climatology of salt transitions and implications for stone weathering. *Sci. Total Environ.* **2011**, *409*, 2577–2585. [CrossRef] [PubMed]

18. Brimblecombe, P.; Grossi, C.M.; Harris, I. Climate change critical to cultural heritage. In *Heritage, Weathering and Conservation*; Fort, R., Álvarez de Buergo, M., Gómez-Heras, C., Vázquez-Calvo, C., Eds.; Taylor & Francis: London, UK, 2006; pp. 387–393.

19. Brimblecombe, P.; Grossi, C.M. Millennium-long damage to building materials in London. *Sci. Total Environ.* **2009**, *407*, 1354–1361. [CrossRef] [PubMed]

20. Gómez-Bolea, A.; Llop, E.; Ariño, X.; Saiz-Jimenez, C.; Bonazza, A.; Messina, P.; Sabbioni, C. Mapping the impact of climate change on biomass accumulation on stone. *J. Cult. Herit.* **2012**, *13*, 254–258. [CrossRef]

21. Sabbioni, C.; Brimblecombe, P.; Cassar, M. *The Atlas of Climate Change Impact on European Cultural Heritage Scientific Analysis and Management Strategies*; Anthem Press: London, UK; New York, NY, USA, 2012; ISBN 9780857282835.

22. Huijbregts, Z.; Kramer, R.P.; Martens, M.H.J.; van Schijndel, A.W.M.; Schellen, H.L. A proposed method to assess the damage risk of future climate change to museum objects in historic buildings. *Build. Environ.* **2012**, *55*, 43–56. [CrossRef]

23. Martens, M.H.J. *Climate Risk Assessment in Museums: Degradation Risks Determined from Temperature and Relative Humidity Data*; Technische Universiteit Eindhoven: Eindhoven, The Netherlands, 2012.

24. Antretter, F.; Kosmann, S.; Kilian, R.; Holm, A.; Ritter, F.; Wehle, B. Controlled Ventilation of Historic Buildings: Assessment of Impact on the Indoor Environment via Hygrothermal Building Simulation. In *Hygrothermal Behavior, Building Pathology and Durability*; de Freitas, V.P., Delgado, J.M.P.Q., Eds.; Springer: Berlin/Heidelberg, Germany, 2013; pp. 93–111, ISBN 978-3-642-31158-1.

25. Kramer, R.; van Schijndel, J.; Schellen, H. Inverse modeling of simplified hygrothermal building models to predict and characterize indoor climates. *Build. Environ.* **2013**, *68*, 87–99. [CrossRef]

26. Leissner, J.; Kaiser, U. *Climate for Culture—Built Cultural Heritage in Times of Climate Change*; Fraunhofer MOEZ: Leipzig, Germany, 2014; ISBN 978-3-00-048328-8.

27. Leissner, J.; Kilian, R.; Kotova, L.; Jacob, D.; Mikolajewicz, U.; Broström, T.; Ashley-Smith, J.; Schellen, H.L.; Martens, M.; Van Schijndel, J.; et al. Climate for culture: Assessing the impact of climate change on the future indoor climate in historic buildings using simulations. *Herit. Sci.* **2015**, *3*, 38. [CrossRef]

28. Köppen, W. Versuch einer Klassifikation der Klimate, vorzugsweise nach ihren Beziehungen zur Pflanzenwelt (Examination of a climate classification preferably according to its relation to the flora). *Geogr. Z.* **1900**, *6*, 593–611, 657–679.

29. Geiger, R. Klassifikation der Klimate nach W. Köppen (Classification of climates after W. Köppen). In *Landolt-Börnstein—Zahlenwerte und Funktionen aus Physik, Chemie, Astronomie, Geophysik und Technik, alte Serie*; Springer: Berlin, Germany, 1954; Volume 3, pp. 603–607.

30. Kottek, M.; Grieser, J.; Beck, C.; Rudolf, B.; Rubel, F. World Map of the Köppen-Geiger climate classification updated. *Meteorol. Z.* **2006**, *15*, 259–263. [CrossRef]

31. Autoridad Nacional del Ambiente. *Atlas Ambiental de la República De Panamá*, 1st ed.; Autoridad Nacional del Ambiente: Panama, Panama, 2010; ISBN 978-9962-651-49-9.

32. Roden, G.I. Sea level variations at Panama. *J. Geophys. Res.* **1963**, *68*, 5701–5710. [CrossRef]

33. Alba Carranza, M.M. *Geografía Descriptiva de la República de Panamá*, 2nd ed.; El Panamá América: Panama, Panama, 1946.

34. Jay, D.A. Evolution of tidal amplitudes in the eastern Pacific Ocean. *Geophys. Res. Lett.* **2009**, *36*, L04603. [CrossRef]

35. Strong, N. Rates and patterns of coastal erosion for the Panama Viejo historical and archeological site. In *Portland Geological Society of America Annual Meeting*; University of Panama: Panama, Panama, 2009; p. 156.

36. Ezcurra, P.; Rivera-Collazo, I.C. An assessment of the impacts of climate change on Puerto Rico's Cultural Heritage with a case study on sea-level rise. *J. Cult. Herit.* **2018**, *32*, 198–209. [CrossRef]

37. Church, J.A.; Clark, P.U.; Cazenave, A.; Gregory, J.M.; Jevrejeva, S.; Levermann, A.; Merrifield, M.A.; Milne, G.A.; Nerem, R.S.; Nunn, P.D.; et al. Sea level change. In *Climate Change 2013: The Physical Science Basis*; Contribution of Working Group I to the Fifth Assessment Report of the Intergovernmental Panel on Climate Change; Stocker, T.F., Qin, D., Plattner, G.-K., Al, E., Eds.; Cambridge University Press: Cambridge, UK; New York, NY, USA, 2013; pp. 1137–1216.

38. Empresa de Transmisión Eléctrica, S.A—ETESA. Available online: http://www.hidromet.com.pa/clima_historicos.php?sensor=2 (accessed on 25 June 2018).

39. Smithsonian Tropical Research Institute—Meteorology and Hydrology Branch, Panama Canal Authority, Republic of Panama. Available online: http://biogeodb.stri.si.edu/physical_monitoring/research/panamacanalauthority (accessed on 25 June 2018).

40. Hazeleger, W.; Severijns, C.; Semmler, T.; Ştefănescu, S.; Yang, S.; Wang, X.; Wyser, K.; Dutra, E.; Baldasano, J.M.; Bintanja, R.; et al. EC-Earth. *Bull. Am. Meteorol. Soc.* **2010**, *91*, 1357–1364. [CrossRef]

41. Hazeleger, W.; Wang, X.; Severijns, C.; Ştefănescu, S.; Bintanja, R.; Sterl, A.; Wyser, K.; Semmler, T.; Yang, S.; van den Hurk, B.; et al. EC-Earth V2.2: Description and validation of a new seamless earth system prediction model. *Clim. Dyn.* **2012**, *39*, 2611–2629. [CrossRef]

42. Davini, P.; von Hardenberg, J.; Corti, S.; Christensen, H.M.; Juricke, S.; Subramanian, A.; Watson, P.A.G.; Weisheimer, A.; Palmer, T.N. Climate SPHINX: Evaluating the impact of resolution and stochastic physics parameterisations in the EC-Earth global climate model. *Geosci. Model Dev.* **2017**, *10*, 1383–1402. [CrossRef]

43. Watson, P.A.G.; Berner, J.; Corti, S.; Davini, P.; von Hardenberg, J.; Sanchez, C.; Weisheimer, A.; Palmer, T.N. The impact of stochastic physics on tropical rainfall variability in global climate models on daily to weekly time scales. *J. Geophys. Res. Atmos.* **2017**, *122*, 5738–5762. [CrossRef]

44. Riahi, K.; Rao, S.; Krey, V.; Cho, C.; Chirkov, V.; Fischer, G.; Kindermann, G.; Nakicenovic, N.; Rafaj, P. RCP 8.5—A scenario of comparatively high greenhouse gas emissions. *Clim. Chang.* **2011**, *109*, 33–57. [CrossRef]

45. Ciantelli, C.; Bonazza, A.; Sabbioni, C.; Suñé Martínez, R.A.; Vaccaro, C. San Fernando Batteries in Portobelo-Panama: Building materials characterization and the environmental impact evaluation. In *Modern Age Fortifications of the Mediterranean Coast—Defensive Architecture of the Mediterranean (Fortmed 2015)*; Editorial Universitat Politècnica de València: Valencia, Spain, 2015.

46. Lipfert, F.W. Atmospheric damage to calcareous stones: Comparison and reconciliation of recent experimental findings. *Atmos. Environ.* **1989**, *23*, 415–429. [CrossRef]

47. Paton, S. *Science of El Niño, The Driest Year Ever?* Trópicos, Magazine of the Smithsonian Tropical Research Institute: Panama, Panama, 2015; pp. 30–33.

48. Camuffo, D. Physical weathering of stones. *Sci. Total Environ.* **1995**, *167*, 1–14. [CrossRef]

49. Gázquez, F.; Rull, F.; Medina, J.; Sanz-Arranz, A.; Sanz, C. Linking groundwater pollution to the decay of 15th-century sculptures in Burgos Cathedral (northern Spain). *Environ. Sci. Pollut. Res.* **2015**, *22*, 15677–15689. [CrossRef] [PubMed]

50. Brimblecombe, P. *Urban Pollution and Changes to Materials and Building Surface*; Air Pollute Imperial College Press: London, UK, 2016; ISBN 978-1-78326-885-6.

51. Ordóñez, S.; La Iglesia, Á.; Louis, M.; García-del-Cura, M.Á. Mineralogical evolution of salt over nine years, after removal of efflorescence and saline crusts from Elche's Old Bridge (Spain). *Constr. Build. Mater.* **2016**, *112*, 343–354. [CrossRef]

52. Satterthwaite, D. *Climate Change and Urbanization: Effects and Implications for Urban Governance*; United Nations Secretariat: New York, NY, USA, 2007.

geosciences

MDPI

Article

Study and Characterization of Environmental Deposition on Marble and Surrogate Substrates at a Monumental Heritage Site

Paola Fermo [1,*], Sara Goidanich [2], Valeria Comite [1], Lucia Toniolo [2] and Davide Gulotta [2]

[1] Dipartimento di Chimica, University of Milan, Via Golgi 19, 20133 Milano, Italy; valeria.comite@unimi.it
[2] Dipartimento di Chimica, Materiali e Ingegneria Chimica "Giulio Natta", Politecnico di Milano, Piazza Leonardo da Vinci 32, 20133 Milano, Italy; sara.goidanich@polimi.it (S.G.); lucia.toniolo@polimi.it (L.T.); davide.gulotta@polimi.it (D.G.)
* Correspondence: paola.fermo@unimi.it; Tel.: +39-338-414-8490

Received: 7 August 2018; Accepted: 10 September 2018; Published: 14 September 2018

Abstract: In this study, the results of the field exposure activity conducted between 2014 and 2017 on the façade of the Milano cathedral (Italy) are reported. The main research aim was to characterize environmental deposition in real exposure conditions and for this purpose, both stone substrates (Candoglia marble) and surrogate substrates (quartz fibre filters) were exposed on the cathedral façade in two sites at different heights. A complete chemical characterization has been performed on quartz filters and marble substrates, i.e., quantification of the deposited aerosol particulate matter (PM) and of the main ions. On quartz filters, the carbonaceous component of deposits was also investigated, as well as the color change induced by soiling, by means of colorimetric measurements. The combined approach exploiting marble and surrogate substrates seems to be a suitable monitoring strategy, although some aspects should be taken into account. In particular, differences in the deposits composition have been highlighted mainly depending on the type of substrate. The environmental data related to atmospheric pollution in Milan for the same period have also been considered but no direct correlations were found between some atmospheric precursors and their related ions in solid deposits.

Keywords: depositions on marble; cultural heritage; exposure tests

1. Introduction

The stone surfaces of the cultural and architectural heritage exposed to the outdoors are subjected to the long-term interaction with the atmospheric pollutants, deriving from both natural and anthropogenic sources [1]. In urban polluted conditions, gas and solid pollutants form complex and multi-component mixtures, which are known to be particularly harmful towards carbonatic substrates [2–5]. The most relevant deterioration processes resulting from such interaction have been extensively studied, in particular, with respect to their chemical reactivity towards the stone substrates [5–8] and to the aesthetic alteration of the exposed surfaces they can trigger, mainly in the form of soiling [9–11].

Efforts have been made to characterize the actual contribution of atmospheric pollutants to damage, in terms of evolution of the deterioration patterns and kinetics of the chemical alteration processes, and monitoring of atmospheric aerosols has also been proposed as a preventive conservation approach for heritage surfaces [12].

This is a particularly challenging task, given the high number of parameters involved, their variability in time and the impact of the currently changing climate scenario. Moreover, considering the specific issue of soiling of the architectural surfaces, deposition rate cannot be estimated only from ambient concentration of pollutants [13] as the inherent features of the substrates, their orientation and the microclimatic aspect are key factors governing the phenomenon. For all these reasons, site exposure tests represent a widely exploited approach to study the response of stone substrates to environmental pressure in actual exposure conditions [10,11,14–17].

In the present work, the results of the field exposure activity conducted between 2014 and 2017 on the main façade of the Milano cathedral are reported. The research was aimed to characterize the environmental deposition in real exposure conditions under the chemical point of view, with a focus on the relationship between atmospheric pollutants and deposition of solid pollutants on the highly valuable architectural surfaces. For this purpose, a specific exposure protocol was designed, combining stone and surrogate substrates. Such approach allowed the comparison between deposition on marble specimens, representative of the actual façade materials, and surrogate substrates provided by quartz fiber filters, which allowed a complete chemical characterization of the particulate matter (PM). The local environmental data corresponding to the two site exposure periods and the historic trends of the main pollutants of Milano city center were also considered.

The Milano cathedral is a remarkable asset of the Italian architectural heritage. Its main façade is entirely made of Candoglia marble, a coarse-grained metamorphic stone with a peculiar white to pink color and presence of grey veins. The façade is cladded by flat slabs and richly decorated with frames, low and high reliefs, architectural elements and sculptures realized between the XVI and XIX centuries. Due to the particularly aggressive exposure conditions of the Milano city center, which caused dramatic damages to the marble surfaces over the years, the façade has been subjected to several extensive conservative intervention during the XX century. The most recent intervention was carried out between 2003 and 2008 and included the general cleaning of all the marble elements, corresponding to a total treated surface of about 10,500 square meters [18]. The results of the field exposure activity here reported are part of the overall monitoring and diagnostic project, supporting the definition of preventive conservation guidelines for the façade.

2. Materials and Methods

Quartz fiber filters (QM-A Whatman, without binder) and Candoglia marble specimens were exposed on-site in outdoor conditions sheltered from direct rain-wash on the façade. Two different exposure sites were identified at 19.70 m (site 1) and 9.40 m (site 2) from the ground level (Figure 1). Site 1 corresponds to the central balcony of the façade. Specimens were exposed on a stainless-steel rack, provided with a PVC roof panel to protect them from the direct rain. All specimens were exposed with a horizontal orientation. Site 2 is located in the south pillar, above the left portal of the façade, exploiting the horizontal surface of the protruding lintel. An overhanging curved tympanum provided shelter from the rain. Over the course of the site exposure, site 2 proved to be particularly protected from most of the rain, whereas in case of particularly intense rain events, some specimens exposed on site 1 were partly in contact with liquid water and therefore were discarded (see Table 1 for the complete list of the analyzed filters). Quartz filters were kept in place by means of plastic containers sealed to the rack or to the lintel surface and leaving the upper surface completely exposed to deposition; marble specimens were just placed on the racks or on the architectural surface, exploiting their inherent higher weight to keep them in a stable position.

Figure 1. Scheme of the façade of the Milano cathedral (courtesy of Veneranda Fabbrica del Duomo di Milano) indicating the location of the exposure sites.

Table 1. List of quartz fiber filters and marble specimens exposed on the Duomo façade in two sites at different height for two sampling periods.

Sample	Exposure Date	Collection	Typology	Sampling Site	Sampling Height
A1					
A2					
A3	July 2014	February 2015			
A4			Quartz fiber filters	SITE 1 Central balcony	19.70 m
A5					
B1					
B2	July 2016	March 2017			
B3					
C1					
C2	July 2014	February 2015			
C3					
C4			Quartz fiber filters	SITE 2 South Pillar	9.40 m
D1					
D2					
D3	July 2016	March 2017			
D4					
D5					
M1					
M2	July 2014	February 2015	Marble specimens	SITE 2 South Pillar	9.40 m
M3					

Quartz filters were pre-conditioned at room temperature in a desiccator for 48 h before weighing; after field exposure, they were again conditioned for 48 h and weighed to determine the deposited powder.

Stone specimens were made of fresh Candoglia marble and were cut into $5 \times 5 \times 2$ cm^3 size. Candoglia marble is a coarse-grained and highly compact marble, with an average porosity below 1%. Its main mineralogical component is calcite, with quartz, pyrite and mica as accessory mineral [19].

The stone substrates were polished with 180 SiC grinding paper to obtain standard reference surfaces (i.e., removal of uneven traces of the cutting procedures) and then washed in deionized water to remove all residues prior to exposure [20].

Two 6-month site exposure were performed in the period July 2014–February 2015 and July 2016–March 2017. A set composed of 5 quartz filters was exposed on sites 1 and 2 for each period, whereas all stone specimens belongs to the first period of exposure (Table 1). At the end of the exposure period, specimens and filters were collected from the sites, stored in sealed plastic containers and transported to the laboratory. A dry removal of the atmospheric deposit from the stone specimens was conducted by means of a stainless-steel scalpel. Particular attention was paid in order to limit the contribution from the stone substrate, i.e., detachment of marble micro-fragments, by working with very mild mechanical action to effectively remove the deposit.

The color change of quartz filters induced by soiling was evaluated by means of colorimetric measurements using a Konica Minolta CM-600D VIS-light spectrophotometer with a D65 illuminant at 8°, wavelength range between 400 nm and 700 nm. At least 3 measurements were conducted on each filter before and after the site exposure. The average data were elaborated according to the CIE L*a*b* standard color system [21]: the variations in lightness and in the saturation of the *b coordinate were calculated respectively as $\Delta L^* = L^*_{after\ exposure} - L^*_{before\ exposure}$, and $\Delta b^* = b^*_{after\ exposure} - b^*_{before\ exposure}$.

A chemical characterization was carried out on both atmospheric deposits removed from the marble specimens and pollutants deposits on quartz fibre filters. Ionic chromatography (IC) was employed for the quantification of the main cationic (Na^+, K^+, Ca^{2+} and Mg^{2+}) and anionic (NO_2^-, NO_3^-, SO_4^{2-} and Cl^-) species by using an ICS-1000 instrument equipped with a conductivity system detector and working with self-regenerating suppressors systems. The solutions suitable for the analysis were prepared according to procedures previously reported [22–24]. Samples of deposits on marble surfaces were characterized by Fourier Transform Infrared Spectroscopy (FTIR) using a Thermo Nicolet 6700 FTIR spectrometer with a DTGS detector in the spectral range 4000–400 cm^{-1}. Samples were prepared in KBr dispersion.

Carbonaceous fraction analysis, i.e., determination of organic carbon (OC) and elemental carbon (EC), has been carried out on filters following the methodology conventionally used for their determination in the aerosol particulate matter [25,26].

The environmental trends of the main pollutants of Milano city center (namely SO_2, NO_x and PM_{10}) have been analyzed to better understand the deposits on both stone and quartz substrates. Environmental data were acquired from the local environmental protection agency (ARPA Lombardia, www.arpalombardia.it) and belong to two sampling sites in the city center, both located within a 500 m linear distance from the façade of the cathedral.

3. Results and Discussion

3.1. Environmental Data and Historic Trends of Some Pollutants of Milano City Center

The presence of sulphur dioxide in the air is due to the combustion of sulphur-containing fossil fuels used mostly for the production of electricity or heating; in urban environment, such as Milan, traces may also be present in automotive emissions that use low refined fuels. In nature, the main source of SO_2 is volcanic activity. Sulphur dioxide is therefore a primary pollutant emitted mostly at the "chimney" level. Moreover, Milan is also subjected to additional stationary sources of SO_2, such as an industrial area, including an oil refinery and a carbon black manufacture sited 40 km West and a refinery sited 50 km South-West of the city. Such sources are sufficiently close to Milan to impact the local atmosphere [27]. From 1970, the technology has made available low-sulphur fuels, the use of which has been imposed by the legislation. Moreover, thanks to the transition of the heating systems to natural gas, the concentrations registered for Milan in the last few years have been further reduced. If the entire data set available for SO_2 is considered [www.arpalombardia.it], starting from 1968, it can be observed that the values have considerably decreased from concentrations of more than 1000 µg/m^3 to much lower values. In particular, in Figure 2, the trend of concentration of the most recent period (2000–2018) is shown.

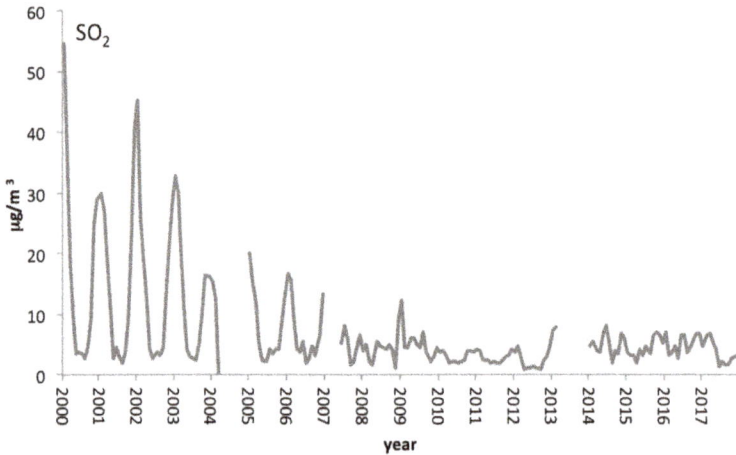

Figure 2. SO_2 concentrations trend registered for Milan monitoring sites (data from ARPA Lombardia) from 2000 until today.

The air concentrations are far below the limits set by Legislative Decree 155/2010 (125 µg/m^3 as daily value). Overall, SO_2 concentrations in Milan are not far from the values observed for other European cities [27].

It is worth noting that in spite of the SO_2 reduction in atmosphere and of the clear improvement of the emission scenario, the presence of sulphate still dominates the chemical composition of the deposits in urban environments [28,29]. This is also confirmed by the results of the chemical characterization of deposits on both filters and marble specimens (as it will be detailed in the next paragraph) after site exposure. This fact could be related to a higher stability and/or a preferential accumulation of sulphate species.

In Figure 3, the trend of NO_x emission during the period of interest for the field exposure tests (2014–2017) has been reported, together with SO_2 concentrations for the same periods. The annual trend of nitrogen dioxide concentrations shows a marked seasonal dependence, with higher values in the winter period, due to both worst dispersive capacity of the atmosphere in the colder months and presence of additional sources, such as domestic heating.

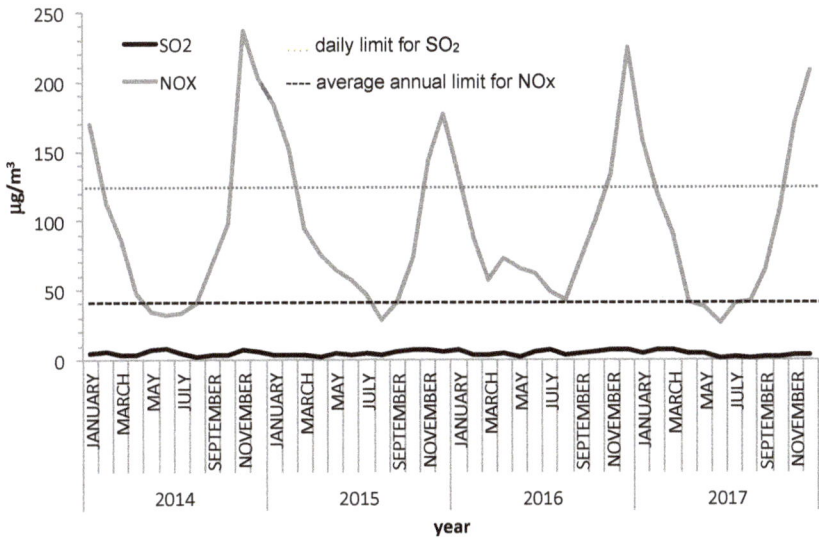

Figure 3. NO_x and SO_2 concentrations trend registered for Milan monitoring sites (data from ARPA Lombardia) during the period of interest for the exposure field tests (2014–2017); daily limit and average annual limit respectively for SO_2 and NO_x (according to D. Lgs. 155/2010) have been reported.

In Figure 4, PM_{10} concentration trend is reported starting from 2002. The same series is not available for $PM_{2.5}$ (that the environmental protection agency started to measure only more recently) but it is worth noting that in Milan $PM_{2.5}/PM_{10}$ ratio is often higher than 0.7 and sometimes almost close to 1, thus indicating that PM10 fraction is mainly composed by $PM_{2.5}$. The analysis of the data collected in the period 2014–2016 [30–32] confirms that particulate matter, for which there are numerous and repeated exceeding of the limits (Figure 4), is one of the most critical parameters for air pollution. It has also a direct impact on cultural heritage surfaces since it is involved in adverse effects such as soiling and sulphation processes [4,6]. More specifically the finer fraction, i.e., $PM_{2.5}$, is involved on both soiling and gypsum nucleation.

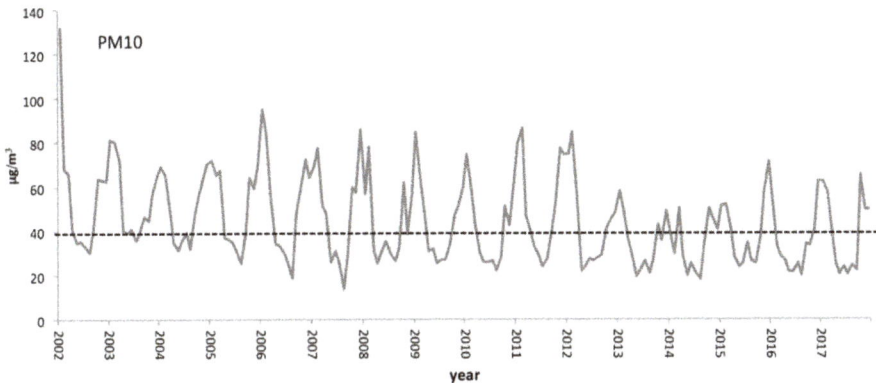

Figure 4. Particulate matter (PM)$_{10}$ concentrations trend registered for Milan monitoring sites (data from ARPA Lombardia); average annual limit (according to D. Lgs. 155/2010) has been reported (dotted line).

Examining the time series with a particular attention to the last decade, it can be noted that PM_{10} concentrations and the number of days of exceeding of the limit on the daily average are appreciably decreased [30–32]. Beyond the meteorological variability, the environmental policies implemented at local, regional and national level contributed to this result, and perhaps, although difficult to quantify, a possible effect linked to the decrease in consumption due to the economic crisis of recent years.

3.2. Characterization of Deposits on Marble and Surrogate Substrates

A preliminary FTIR (Fourier Transform Infrared Spectroscopy) characterization has been carried to assess the general composition of the deposits on marble substrates. FTIR analysis of samples of deposits from specimens exposed in site 2 indicates a prevailing calcitic composition (a representative FTIR spectrum is reported in Figure 5). Calcite is identified by the characteristic absorption peaks at 1425, 875 and 715 cm^{-1}. A significant presence of calcium sulphate, as gypsum, can also be observed (peals at 3541–3407, 1620, 1143–1115 and 670–602 cm^{-1}), together with nitrates (peak at 1385 cm^{-1}). Peak at 1036 cm^{-1} (stretching mode of the Si-O bond) and the doublet at 797–779 cm^{-1} are due to the silicatic fraction of the deposit and to the presence of crystalline quartz, respectively. Moreover, the presence of organic compounds within the deposit is also indicated by the absorption peaks in the 2850–2920 cm^{-1} region.

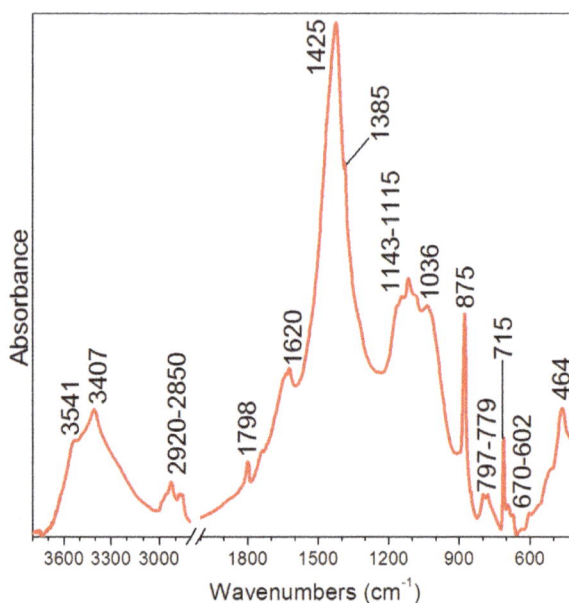

Figure 5. FTIR characterization of a representative deposit on stone specimen (site 2).

The main ionic species determined for each quartz filter and for the deposits on marble specimens, together with the average values for each site and for the two exposure periods, are reported in Table 2. In Figures 6 and 7 the average concentrations of anions and cations for quartz filter are shown, while in Figure 8 anions and cations average concentrations determined in the deposits on marble are reported. It has to be pointed out that the ionic composition of deposits from marble might be slightly affected by a contribution belonging to the substrate itself, although sampling was performed by mild and controlled mechanical actions.

Table 2. Main ionic species determined on quartz fiber filters and powder deposits on marble specimens exposed in the two sites on the Duomo façade for two periods.

		$\mu g/cm^2$							
		Anions				Cations			
		Cl^-	NO_3^-	SO_4^{2-}	Na^+	NH_4^+	K^+	Mg^{2+}	Ca^{2+}
		July 2014–February 2015							
Site 1 Central terrace 19.70 m	A1	7.0	19.7	28.2	7.2	1.1	3.0	1.9	39.0
	A2	5.3	10.2	18.1	5.5	n.a.	1.4	1.3	38.1
	A3	5.7	14.5	24.7	4.9	0.1	1.6	1.1	16.3
	A4	5.9	15.0	26.3	4.5	0.1	1.4	1.1	15.2
	A5	4.7	4.8	18.5	4.0	0.1	0.9	0.7	11.5
	Average	5.7	12.9	23.1	5.2	0.3	1.7	1.2	24.0
	Std dev	0.4	2.5	2.1	0.6	0.2	0.4	0.2	6.0
		July 2016–March 2017							
	B1	17.6	21.9	23.0	14.9	0.4	2.7	n.a	64.2
	B2	17.6	6.6	20.4	13.7	0.3	1.9	n.a	54.5
	B3	16.7	17.2	25.9	13.8	0.4	2.2	n.a	58.9
	Average	17.3	15.2	23.1	14.1	0.4	2.3	n.a	59.2
	Std dev	0.3	4.6	1.6	0.4	0.0	0.3	n.a	2.8
		July 2014–February 2015							
Site 2 South Pillar 9.40 m	C1	4.0	2.5	17.1	2.9	0.1	0.8	0.7	11.9
	C2	4.3	7.2	13.6	3.0	0.1	1.1	0.8	13.6
	C3	3.8	6.8	11.9	2.9	0.1	0.9	0.8	14.3
	C4	4.1	6.7	12.3	3.2	0.1	1.0	0.8	15.0
	Average	4.0	5.8	13.7	3.0	0.1	0.9	0.8	13.7
	Std dev	0.1	1.1	1.2	0.1	0.0	0.1	0.0	0.7
		July 2016–March 2017							
Site 2 South Pillar 9.40 m	D1	11.8	7.0	7.8	9.7	0.4	0.9	n.a	30.8
	D2	15.6	7.3	8.9	11.5	0.4	1.1	n.a	33.1
	D3	14.0	7.8	9.6	9.3	0.3	0.9	n.a	34.8
	D4	17.3	9.0	9.7	11.0	0.3	1.0	n.a	36.7
	D5	15.5	6.7	9.7	11.2	0.2	1.0	n.a	38.2
	Average	14.9	7.5	9.1	10.5	0.3	1.0	n.a	34.7
	Std dev	0.9	0.4	0.4	0.4	0.0	0.0	n.a	1.3
		July 2014–February 2015							
Site 2 South Pillar 9.40 m	M1	1.3	1.9	25.4	2.0	0.0	0.7	0.4	18.9
	M2	1.4	2.7	23.8	1.4	0.6	0.7	0.3	17.9
	M3	1.2	5.9	20.6	1.2	n.a	0.7	0.3	16.1
	Average	1.3	3.5	23.3	1.5	0.3	0.7	0.3	17.6
	Std dev	0.1	1.2	1.4	0.3	0.2	0.0	0.0	0.8

Sulphate is the anionic species present with the highest concentration in both filters and marble specimens, in accordance with what typically observed for aerosol particulate matter in urban conditions [28,29], with the exception of the filters exposed in site 2 in the second exposure campaign (2016–2017), during which the main anionic species was chloride. It is worth noting that for marble specimens SO_4^{2-} concentration is almost twice the one of the filters. In this context, the contribution of an early stage sulphation process due to the interaction between atmospheric SO_2 and the outermost portion of the stone substrate cannot be excluded.

Such process was identified as an important decay mechanism of the façade in the past [33] and can be favored by the specific exposure conditions [2]. Furthermore, the higher sulphate/nitrate ratio found for the marble surfaces with respect to quartz filters is in accordance with what already observed in a previous study [17]. The overall characterization results indicate that the deposits have a significant potential for deterioration of the stone surfaces. In particular, high amount of sulphates, soluble salts and silicatic compounds are typically associated to damage layers in urban conditions [34].

As regards the differences between the two exposure sites, during both periods sulphates are always higher in site 1 with respect to site 2 (Figure 6). The same trend is observable for the other anions. The compositional differences between the two sites are influenced by several factors and are not easy to be explained. In particular, besides the different height, the complex geometry of the façade's elements surely plays a role in air circulation [35], together with the presence of urban canyons effects [36], thus affecting particles deposition.

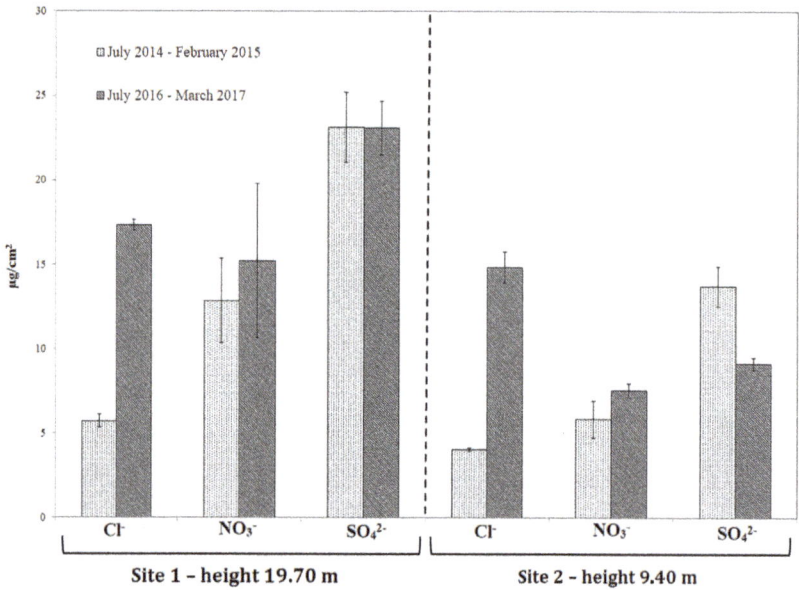

Figure 6. Anions average concentrations determined on the quartz fiber filters exposed in the two sites on the Duomo façade during the two periods.

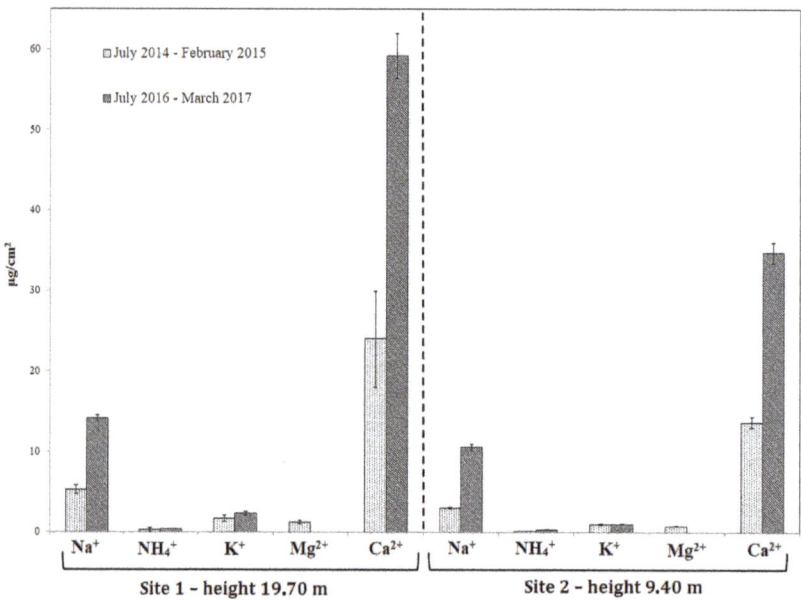

Figure 7. Cations average concentrations determined on the quartz fiber filters exposed in the two sites on the Duomo façade during the two periods.

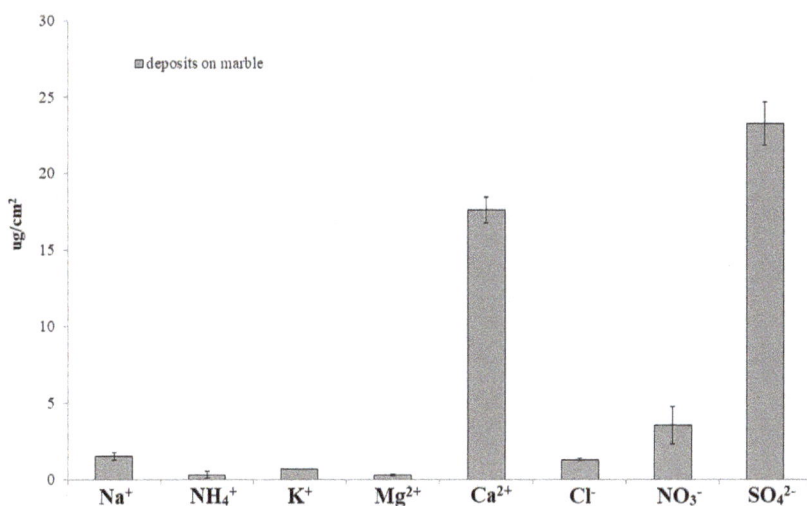

Figure 8. Cations and anions average concentrations determined in the deposits on marble specimens exposed on the Duomo façade during the first period (2014–2015).

In general, a very low concentration of NH_{4+} was detected upon exposure (Figure 7), while it is known that on filters SO_4^{2-} is typically found as ammonium sulphate [23]. The low concentration detected may be due to the decomposition of such compound. Since sulphate in the aerosol particulate matter could be of both anthropogenic or marine origin, an apportionment between these two sources has been calculated in accordance with what reported in the literature [37–41]. This approach is based on the fact that sulphate of marine origin is present as sodium sulphate and, as a consequence, sulphate of anthropogenic origin ($NSS\_SO_4^{2-}$ = non-sea salt sulphate) can be calculated from total sulphate by subtracting the contribution from Na_2SO_4 ($SS\_SO_4^{2-}$ = sea salt sulphate), considering the ratio between these two elements in sea salt. As it can be observed in Figure 9, during the first campaign (2014–2015) the calculated contribution from sea salt is lower with respect to the second campaign (2016–2017) and this is also in accordance with the lower chloride concentrations detected in 2014–2015 (Figure 6).

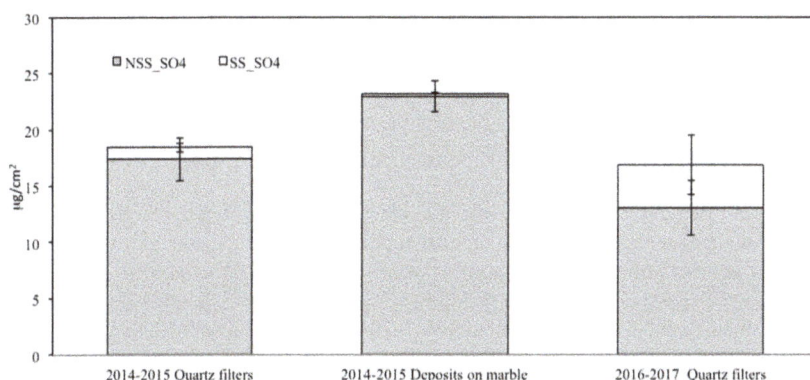

Figure 9. $NSS\_SO_4^{2-}$ (non-sea salt sulphate) and $SS\_SO_4^{2-}$ (sea salt sulphate) contributions determined for quartz fiber filter and deposits on marble specimens in the two exposure periods.

To support this evaluation, the contribution of sea salt to Cl^- concentration has also been calculated for both filters and deposits on marble (Table 3) and compared to the characteristic elemental ratios reported for chloride of marine origin [37,38,41]. Cl/Na ratios lower than the reference value of 1.8, typical of sea salt, indicate the presence of other sources of Na, in addition to marine spray, such as re-suspended dust.

Table 3. Elemental ratios of interest for the determination of chloride contribution of marine origin calculated for quartz fiber filters and marble specimens exposed on the Duomo façade during the two periods.

	Cl/Na	K/Na	Mg/Na	Ca/Na	SO_4^{2-}/Na
Marine Origin	1.80	0.04	0.12	0.04	0.25
Deposits on Marble July 2014–February 2015	1.12	0.59	0.25	15.21	20.08
Filter Deposits July 2014–February 2015	1.19	0.32	0.24	4.59	4.49
Filter Deposits July 2016–March 2017	1.30	0.13	0.00	3.81	1.31

Furthermore, SO_4^{2-}/Na ratios higher than the reference value for sea salt (i.e., 0.25), confirm the predominantly anthropogenic origin of sulphates (Table 3). It is also quite interesting comparing SO_4^{2-}/Cl ratios found in the quartz filters in this study with the typical ratios found in the aerosol particulate matter in the Po Valley [42]. In the case of PM_{10} this ratio ranges from 3.75 to 5 [27] while in the present study SO_4^{2-}/Cl ranges from 1 to 3.78. Such lower values arise from a slightly higher contribution from chloride. This is in accordance with the fact that in the present study the fraction collected on filters is the total suspended particulate matter and chlorides are generally more concentrated in the coarser particles. A possible additional source of sodium chloride during winter may also be due to de-icing treatments.

As regards cations concentrations (Figure 7), calcium is the main species. The higher calcium values registered for the second sampling period could be attributed to dust transport phenomena, which are not unusual in Milan.

Finally, a generally lower concentration of nitrates has been found in the deposits on marble specimens with respect to the quartz filters. Considering the high solubility of nitrates, their concentration on the marble surfaces can be affected by dissolution phenomena and consequently partial migration within the low porous microstructure of the specimens can occur.

In terms of PM deposition rates (DR, calculated as $\mu g/cm^2$ month^{-1}), the values obtained in the present study for quartz filters are in agreement with those obtained for marble surfaces in a study recently carried out in Milan [17]: on average 68.9 $\mu g/cm^2$ month^{-1} for site 1 and 54.1 $\mu g/cm^2$ month^{-1} for site 2 against values in the range of about 50–70 $\mu g/cm^2$ month^{-1} measured in Milan for Carrara marble and quartz filters as surrogate surfaces [17] during the same period of the first exposure test (2014–2015).

As regards marble specimens, DR was 73.2 $\mu g/cm^2$ month^{-1}, slightly higher than the value measured for the quartz filters exposed in parallel in the same site (site 2), i.e., 54.1 $\mu g/cm^2$ month^{-1}, indicating that DR on Candoglia marble is higher with respect to the filter of about 35%.

From the comparison of the overall results of deposits on filters and on marble specimens during the first exposure campaign some considerations can be drawn. As for the deposition rate, the use of filters may lead to an underestimation of the deposition. The same applies for the sulphates, which in the case of marble substrates are generally found in higher concentration with respect to the deposits on filters. As previously discussed, this can also be related to the inherent reactivity of the stone material. As far as the nitrates content are concerned, the generally lower values found on marble specimens can be related to decomposition or solubilization mechanisms. In this case, the use of filter may allow a more reliable estimation of the actual content of these species.

In Table 4 OC, and TC concentrations determined for the quartz fiber filters are reported together with their average values. EC values have not been reported since they were below the detection limit of the technique [23]. Due to the very low elemental carbon concentrations OC and TC values are indeed very similar. In Figure 10 average OC, TC and particulate matter concentrations on filters are shown. For both sites higher concentrations have been registered for the second exposure period with the greatest difference detected for site 1.

Table 4. OC (organic carbon) and TC (total carbon) determined on quartz fiber filters and powder deposits exposed in the two sites on the Duomo façade during the two periods.

		$\mu g/cm^2$	
		OC	TC
	July 2014–February 2015		
	A1	52.67	52.71
	A2	56.05	56.10
	A3	58.05	58.09
	A4	52.46	52.49
Site 1 Central	Average	54.81	54.85
terrace 19.70 m	Std dev	1.42	1.42
	July 2016–March 2017		
	B1	73.51	73.56
	B2	77.37	77.41
	B3	78.74	78.77
	Average	76.54	76.58
	Std dev	1.57	1.56
	July 2014–February 2015		
	C1	46.46	46.49
	C2	48.27	48.31
	C3	48.43	48.48
	C4	44.96	45.00
	Average	47.03	47.07
Site 2 South Pillar	Std dev	0.82	0.82
9.40 m	July 2016–March 2017		
	D1	58.14	58.20
	D2	59.41	59.46
	D3	55.44	55.51
	D4	58.79	58.86
	D5	59.58	59.64
	Average	58.27	58.33
	Std dev	0.67	0.67

A good correlation was found between the trend observed for the carbon fraction concentrations (both OC and TC) and the color variations measured on filters (Figure 11). The variation of the L* coordinate (ΔL*) corresponds to the progressively reduced lightness resulting from particulate matter accumulation, which is also associated to an increase of the b* coordinate (Δb*), due to a slight yellowing effect. After six months of exposure, a significant blackening effect can be detected in all cases, with ΔL* values always greater than 11 units. The yellowing effect is less evident and generally ranges between 3.3 (site 2, second exposure period) and 5.3 units (site 1, second exposure period). Moreover, as observed for the carbon fraction trends, the entity of L* and b* variations are more intense during the second exposure period, especially on filters exposed in site 1. In the present study, as already mentioned, EC values were below the limit of detection and consequently the relationship between this specific parameter and the surface color variation cannot be highlighted. Nevertheless, it was possible to observe a relation between the carbonaceous component and the color variation.

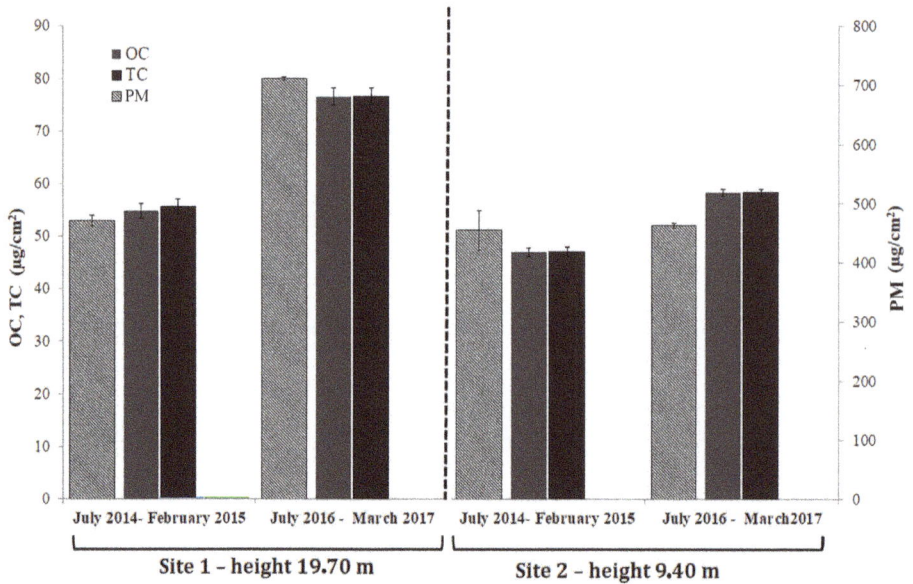

Figure 10. OC (organic carbon), TC (total carbon) and PM weights determined on quartz fiber filters exposed in the two sites on the Duomo façade for two periods.

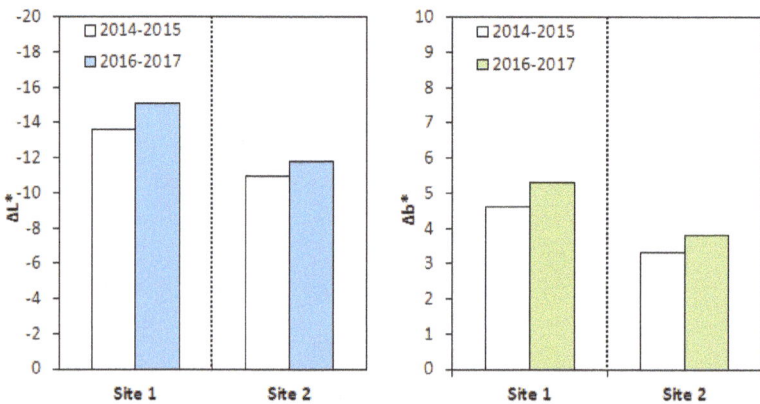

Figure 11. Colorimetric variation of exposed filters: variation of the lightness (ΔL^*, **left**) and of the saturation of the yellow coordinate (Δb^*, **right**) during the two exposure periods.

In Figure 12 the average concentrations of the major solid and gaseous pollutants (data elaborated from www.arpalombardia.it) are reported for the two periods of interest for the present study, while in Figure 13 nitrate and sulphate average concentrations in deposits on quartz filter exposed on the façade and taking into account both sites are reported. The standard deviations of atmospheric data indicate that the average concentrations of atmospheric pollutants in the two considered periods is rather comparable. A direct correlation between the concentrations of the two precursor (i.e., NO_x and SO_2) and the corresponding ions cannot be evidenced: in particular, the remarkable predominance of

NO_x over SO_2 in atmosphere is not associated to deposits enriched in nitrates. However, it is worth noting that the wet or dry deposition on surfaces is a very complex phenomenon where numerous variables (including, for example, relative humidity) are involved.

Notwithstanding the significant reduction of atmospheric SO_2 concentration (Figure 2), the results of the present study show that sulphates are still the main anionic constituents of the deposits on both filters and marble substrates. This is in accordance with studies showing that significant formation of sulphates, as gypsum, can occur even in areas characterized with rather low levels of atmospheric SO_2 [7].

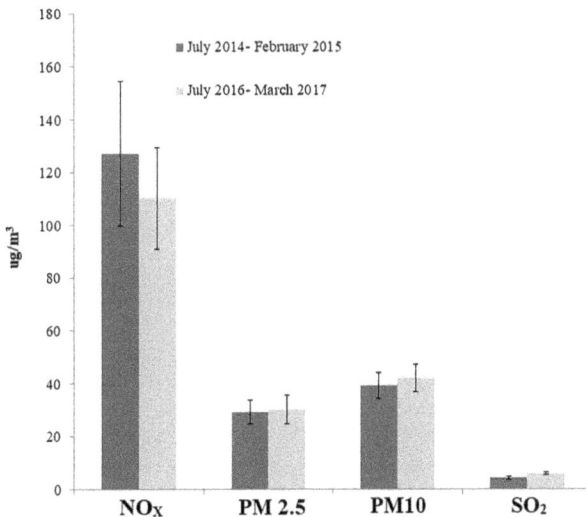

Figure 12. Main pollutants average concentrations for Milan monitoring sites (data from ARPA Lombardia) for the two periods of interest for the exposure field tests.

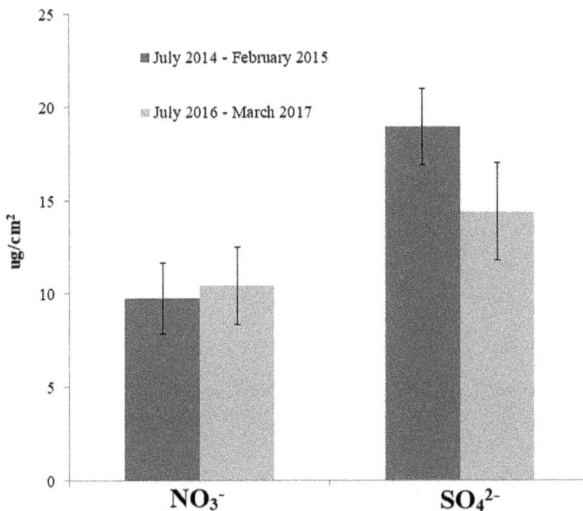

Figure 13. Sulphate and nitrate average concentrations determined on quartz fiber filters for the two periods of interest for the exposure field tests (data calculated as averages between the two sites).

4. Conclusions

In the present research, site exposure tests were carried out between 2014 and 2017 on the main façade of the Milano cathedral for the chemical characterization of the environmental deposition in real exposure conditions (including the quantifications of the main anions and the carbonaceous components). The combined approach exploiting marble and surrogate substrate seems to be a suitable monitoring strategy, although some aspects needs to be pointed out.

As far as PM deposition rate (DR) and sulphates content are considered, it can be observed that in the same site (site 2) they were found to be higher for marble surfaces than for quartz filters. Nitrates, on the other hand, showed an opposite trend. These aspects require further investigation over additional monitoring periods and exposure sites in order to confirm the different response of the stone specimens with respect to the filters. Considering the relevant role of sulphates in stone deterioration, their possible underestimation should be considered if monitoring is conducted by means of filters. Overall, the precise evaluation of DR and deposits composition is fundamental for the definition of effective conservation and mitigation strategies in a rapidly changing environment.

Deposit compositions were compared to the environmental data calculated for the Milan city centre over the same exposure periods. It has to be pointed out that the methodologies of acquisition of environmental data (i.e., based on active atmospheric samplers) are rather different from the passive approach followed in the present work for deposits collection. No direct correlation was found between some atmospheric precursors and their related ions in solid deposits. In particular, a rather stable concentration of environmental nitrogen and sulphur based gas pollutants during the two site exposures was associated to remarkable variations of ions content of the deposits collected on filters in the same periods. This confirms that concentrations of the main pollutants in atmosphere, although relevant in the evaluation of the overall environmental harmfulness, cannot be directly used as proxies to assess the nature of deposits on architectural surfaces exposed in urban contexts.

Considering the historic data of gas pollutants, it was observed that the deposition of sulphates remains consistent despite the significant decreases in atmospheric SO_2 recorded in recent years. According to the trends measured by the regional environmental protection agency, PM concentration, on the other hand, does not seem to show a clear decrease. Given its correlation with the deposition of carbonaceous particles, thus affecting soiling, its actual role on exposed surfaces should be further monitored.

Author Contributions: Conceptualization and methodology, P.F., S.G., D.G. and L.T.; Investigation, V.C., P.F. and D.G.; Visualization, V.C., P.F., S.G. and D.G.; Writing-original draft preparation, P.F. and D.G.; Writing-review and editing, P.F., S.G. and D.G.

Funding: This research has been carried out in the framework of the project "La Facciata del Duomo di Milano. Piano di monitoraggio e conservazione programmata a seguito dell'intervento straordinario di restauro" funded by Veneranda Fabbrica del Duomo di Milano.

Acknowledgments: The authors wish to acknowledge Veneranda Fabbrica del Duomo di Milano and Ing. Francesco Canali for the support and collaboration during the activity.

Conflicts of Interest: The authors declare no conflicts of interest.

References

1. Watt, J.; Tidblad, J.; Kucera, V.; Hamilton, R. *The Effect of Air Pollution on Cultural Heritage*; Watt, J., Ed.; Springer-Verlag: Berlin, Germany, 2009; ISBN 978-0-387-84892-1.
2. Camuffo, D.; Del Monte, M.; Sabbioni, C. Origin and growth mechanisms of the sulfated crusts on urban limestone. *Water Air Soil Pollut.* **1982**, *19*, 351–359. [CrossRef]
3. Del Monte, M.; Sabbioni, C.; Vittori, O. Urban stone sulphation and oil-fired carbonaceous particles. *Sci. Total Environ.* **1984**, *36*, 369–376. [CrossRef]
4. Sabbioni, C. Contribution of atmospheric deposition to the formation of damage layers. *Sci. Total Environ.* **1995**, *167*, 49–55. [CrossRef]

5. Steiger, M.; Charola, A.E.; Sterflinger, K. Weathering and deterioration. In *Stone in Architecture*; Siegesmund, S., Snethlage, R., Eds.; Springer: Berlin, Germany, 2011; pp. 225–313.
6. Rodriguez-Navarro, C.; Sebastian, E. Role of particulate matter from vehicle exhaust on porous building stones (limestone) sulfation. *Sci. Total Environ.* **1996**, *187*, 79–91. [CrossRef]
7. Török, A.; Licha, T.; Simon, K.; Siegesmund, S. Urban and rural limestone weathering; the contribution of dust to black crust formation. *Environ. Earth Sci.* **2011**, *63*, 675–693. [CrossRef]
8. Gibeaux, S.; Vázquez, P.; Kock, T.; De Cnudde, V.; Thomachot-Schneider, C. Weathering assessment under X-ray tomography of building stones exposed to acid atmospheres at current pollution rate. *Constr. Build. Mater.* **2018**, *168*, 187–198. [CrossRef]
9. Viles, H.A.; Taylor, M.P.; Yates, T.J.S.; Massey, S.W. Soiling and decay of N. M. E. P. limestone tablets. *Sci. Total Environ.* **2002**, *292*, 215–229. [CrossRef]
10. Grossi, C.M.; Esbert, R.M.; Dã, F.; Alonso, F.J. Soiling of building stones in urban environments. *Build. Environ.* **2003**, *38*, 147–159. [CrossRef]
11. Urosevic, M.; Yebra-Rodríguez, A.; Sebastián-Pardo, E.; Cardell, C. Black soiling of an architectural limestone during two-year term exposure to urban air in the city of Granada (S Spain). *Sci. Total Environ.* **2012**, *414*, 564–575. [CrossRef] [PubMed]
12. Ghedini, N.; Ozga, I.; Bonazza, A.; Dilillo, M.; Cachier, H.; Sabbioni, C. Atmospheric aerosol monitoring as a strategy for the preventive conservation of urban monumental heritage: The Florence Baptistery. *Atmos. Environ.* **2011**, *45*, 5979–5987. [CrossRef]
13. Watt, J.; Tidblad, J.; Kucera, V.; Hamilton, R. Monitoring, modelling and mapping. In *The Effects of Air Pollution on Cultural Heritage*; Hamilton, R., Kucera, V., Tidblad, J., Watt, J., Eds.; Springer: Berlin, Germany, 2009; pp. 29–51.
14. Zappia, G.; Sabbioni, C.; Riontino, C.; Gobbi, G.; Favoni, O. Exposure tests of building materials in urban atmosphere. *Sci. Total Environ.* **1998**, *224*, 235–244. [CrossRef]
15. Viles, H.A.; Gorbushina, A.A. Soiling and microbial colonisation on urban roadside limestone: A three year study in Oxford, England. *Build. Environ.* **2003**, *38*, 1217–1224. [CrossRef]
16. Comite, V.; De Buergo, M.Á.; Barca, D.; Belfiore, C.M.; Bonazza, A.; La Russa, M.F.; Pezzino, A.; Randazzo, L.; Ruffolo, S.A. Damage monitoring on carbonate stones: Field exposure tests contributing to pollution impact evaluation in two Italian sites. *Constr. Build. Mater.* **2017**, *152*, 907–922. [CrossRef]
17. Ferrero, L.; Casati, M.; Nobili, L.; Angelo, L.D.; Rovelli, G.; Sangiorgi, G.; Rizzi, C.; Perrone, M.G.; Sansonetti, A.; Conti, C.; et al. Chemically and size-resolved particulate matter dry deposition on stone and surrogate surfaces inside and outside the low emission zone of Milan: Application of a newly developed Deposition Box. *Environ. Sci. Pollut. Res.* **2018**, *25*, 9402–9415. [CrossRef] [PubMed]
18. Morlin Visconti Castiglione, B. Alcune note sui lavori di restauro della facciata del Duomo, 2003–2008. In *Nuovi Annali—Rassegna di Studi e Contributi per il Duomo di Milano*; Edizioni Et: Milano, Italy, 2008.
19. Toniolo, L.; Zerbi, C.M.; Bugini, R. Black layers on historical architecture. *Environ. Sci. Pollut. Res.* **2009**, *16*, 218–226. [CrossRef] [PubMed]
20. UNI 10859. *Cultural Heritage—Natural Stones—Determination of Water Absorption by Capillarity*; Ente Nazionale Italiano di Unificazione: Milan, Italy, 2000.
21. UNI-EN 15886. *Conservation of Cultural Property—Test Methods—Colour Measurement of Surfaces*; Ente Nazionale Italiano di Unificazione: Milan, Italy, 2010.
22. Fermo, P.; Turrion, R.G.; Rosa, M.; Omegna, A. A new approach to assess the chemical composition of powder deposits damaging the stone surfaces of historical monuments. *Environ. Sci. Pollut. Res.* **2015**, *22*, 6262–6270. [CrossRef] [PubMed]
23. Piazzalunga, A.; Bernardoni, V.; Fermo, P.; Vecchi, R. Optimisation of analytical procedures for the quantification of ionic and carbonaceous fractions in the atmospheric aerosol and applications to ambient samples. *Anal. Bioanal. Chem.* **2013**, *405*, 1123–1132. [CrossRef] [PubMed]
24. La Russa, M.F.; Fermo, P.; Comite, V.; Belfiore, C.M.; Barca, D.; Cerioni, A.; De Santis, M.; Barbagallo, L.F.; Ricca, M.; Ruffolo, S.A. The Oceanus statue of the Fontana di Trevi (Rome): The analysis of black crust as a tool to investigate the urban air pollution and its impact on the stone degradation. *Sci. Total Environ.* **2017**, *593*, 297–309. [CrossRef] [PubMed]
25. Fermo, P.; Piazzalunga, A.; Vecchi, R.; Valli, G.; Ceriani, M. A TGA/FT-IR study for measuring OC and EC in aerosol samples. *Atmos. Chem. Phys.* **2006**, *6*, 255–266. [CrossRef]

26. Panteliadis, P.; Hafkenscheid, T.; Cary, B.; Diapouli, E.; Fischer, A.; Favez, O.; Quincey, P.; Viana, M.; Hitzenberger, R.; Vecchi, R.; et al. ECOC comparison exercise with identical thermal protocols after temperature offset correction—Instrument diagnostics by in-depth evaluation of operational parameters. *Atmos. Meas. Tech.* **2015**, *8*, 779–792. [CrossRef]

27. Bigi, A.; Bianchi, F.; De Gennaro, G.; Di Gilio, A.; Fermo, P.; Ghermandi, G.; Prévôt, A.S.H.; Urbani, M.; Valli, G.; Vecchi, R.; et al. Hourly composition of gas and particle phase pollutants at a central urban background site in Milan, Italy. *Atmos. Res.* **2017**, *186*, 83–94. [CrossRef]

28. Bernardoni, V.; Elser, M.; Valli, G.; Valentini, S.; Bigi, A.; Fermo, P.; Piazzalunga, A.; Vecchi, R. Size-segregated aerosol in a hot-spot pollution urban area: Chemical composition and three-way source apportionment. *Environ. Pollut.* **2017**, *231*, 601–611. [CrossRef] [PubMed]

29. Bernardoni, V.; Vecchi, R.; Valli, G.; Piazzalunga, A.; Fermo, P. PM_{10} source apportionment in Milan (Italy) using time-resolved data. *Sci. Total Environ.* **2011**, *409*, 4788–4795. [CrossRef] [PubMed]

30. Algieri, A.; Chiesa, M.; Cigolini, G.; Colombi, C.; Cosenza, R.; Cuccia, E.; Dal Santo, U.; Dal Zotto, M.; Ferrari, R.; Gentile, N.; et al. *Rapporto Sulla Qualità Dell' Aria Della Città Metropolitana di Milano*; ARPA Lombardia: Milan, Italy, 2014.

31. Algieri, A.; Chiesa, M.; Cigolini, G.; Colombi, C.; Cosenza, R.; Cuccia, E.; Dal Santo, U.; Dal Zotto, M.; Ferrari, R.; Gentile, N.; et al. *Rapporto Sulla Qualità Dell' Aria Della Città Metropolitana di Milano*; ARPA Lombardia: Milan, Italy, 2015. Available online: http://www.arpalombardia.it/qariafiles/RelazioniAnnuali/RQA_MI_2015.pdf (accessed on 14 September 2018).

32. Algieri, A.; Chiesa, M.; Cigolini, G.; Colombi, C.; Cosenza, R.; Cuccia, E.; Dal Santo, U.; Dal Zotto, M.; Ferrari, R.; Gentile, N.; et al. *Rapporto Sulla Qualità Dell' Aria Della Città Metropolitana di Milano*; ARPA Lombardia: Milan, Italy, 2016. Available online: http://www.arpalombardia.it/qariafiles/RelazioniAnnuali/RQA_MI_2016.pdf (accessed on 14 September 2018).

33. Pedrazzani, R.; Alessandri, I.; Bontempi, E.; Cappitelli, F.; Cianci, M.; Pantos, E.; Toniolo, L.; Depero, L.E. Study of sulphation of Candoglia marble by means of micro X-ray diffraction experiments. *Appl. Phys. A Mater. Sci. Process.* **2006**, *83*, 689–694. [CrossRef]

34. Bonazza, A.; Sabbioni, C.; Ghedini, N. Quantitative data on carbon fractions in interpretation of black crust and soiling on European built heritage. *Atmos. Environ.* **2005**, *39*, 2607–2608. [CrossRef]

35. Auras, M.; Bundschuh, P.; Eichhorn, J.; Kirchner, D.; Mach, M.; Seewald, B.; Snethlage, R. Salt deposition and soiling of stone facades by traffic-induced immissions. *Environ. Earth Sci.* **2018**, *77*, 1–16. [CrossRef]

36. Quang, T.N.; He, C.; Morawska, L.; Knibbs, L.D.; Falk, M. Vertical particle concentration profiles around urban office buildings. *Atmos. Chem. Phys.* **2012**, *12*, 5017–5030. [CrossRef]

37. Karthikeyan, S.; Balasubramanian, R. Determination of water-soluble inorganic and organic species in atmospheric fine particulate matter. *Microchem. J.* **2006**, *82*, 49–55. [CrossRef]

38. Keene, W.C.; Pszenny, A.A.P.; Galloway, J.N.; Hawley, M.E. Sea-salt corrections and interpretation of constitutent ratios in marine precipitation. *J. Geophys. Res.* **1986**, *91*, 6647–6658. [CrossRef]

39. La Russa, M.F.; Comite, V.; Aly, N.; Barca, D.; Fermo, P.; Rovella, N.; Antonelli, F.; Tesser, E.; Aquino, M.; Ruffolo, S.A. Black crusts on Venetian built heritage, investigation on the impact of pollution sources on their composition. *Eur. Phys. J. Plus* **2018**, *133*, 370. [CrossRef]

40. Atzei, D.; Fermo, P.; Vecchi, R.; Fantauzzi, M.; Comite, V.; Valli, G. Composition and origin of $PM_{2.5}$ in Mediterranean countryside. *Environ. Pollut.* **2018**, in submission.

41. Seinfeld, J.H.; Pandis, S.N. *Atmospheric Chemistry and Physics*; Wiley: New York, NY, USA, 1998.

42. Colombi, C. Relazione Campagna PoAIR Febbraio 2014. Available online: http://www.arpalombardia.it/qariafiles/varie/Relazione%20campagna%20PoAIR%20febbraio%202014.pdf (accessed on 14 September 2018).

geosciences

MDPI

Article

Estimators of the Impact of Climate Change in Salt Weathering of Cultural Heritage

Beatriz Menéndez

Géosciences et Environnement Cergy—Université de Cergy-Pontoise, 95000 Neuville sur Oise, France;
beatriz.menendez@u-cergy.fr; Tel.: +33-134-257-362

Received: 28 September 2018; Accepted: 31 October 2018; Published: 3 November 2018

Abstract: Changes induced by climate change in salt weathering of built cultural heritage are estimated in different ways, but generally as a function of phase changes phenomena of two common salts, sodium chloride and sodium sulfate. We propose to use not only these salts, but also other common salts as calcium sulfate, or mixtures of chlorides, sulfates, and nitrates of sodium, calcium, magnesium, and potassium. Comparisons between the predicted changes in salt weathering obtained for single salts and for combinations of different salts are presented. We applied the proposed methodology to 41 locations uniformly distributed in France. The results show that estimations of actual and evolution of future weathering depend on the selected salt or combination of salts. According to our results, when using a combination of different salts, weathering evolution is less favorable (more damage in the future) than when using a single salt.

Keywords: salt climatology; climate change; built cultural heritage weathering

---

1. Introduction

Salts are almost omnipresent in built cultural heritage. Their nature (chemical composition and "mineralogical" phase), abundance and location in the building depend on the stone composition, wall orientation, and environmental parameters. Most of these salts are the result of the interaction between building stones and external agents (sea salt, atmospheric pollution, and underground water) or other construction materials: (mortars, plasters). In very specific cases, the building is made of evaporite rocks, as in the Old Shali town in Egypt [1]. In such a case, the environmental conditions will strongly control the durability of the cultural building. Salt weathering mechanisms are very complex and are not completely understood. Several papers describing salt weathering mechanisms have been published, as, for example, the paper of Steiger (2003) [2]. Pollutants deposition takes place by two different mechanism, dry deposition, and wet deposition. [3]. These pollutants, dissolved in water, will react with the stone minerals to form soluble salts [2]. Salt get into the masonry dissolved in water by two principal ways, capillarity rise and infiltration of rainwater [4].

Depending on the environmental conditions, mainly temperature and relative humidity, different salts may crystallize and they will do it at different locations inside the stone or at the surface of the wall. Arnold and Zehnder [5] propose a model of salt distribution in the masonry wall as a function of the height. In the lower area (A), weathering is less important than in the next upper zone (B) where a group of weakly to moderate soluble salts is responsible of an important weathering. Next upper area (C) is generally damp and dark and shows an accumulation of more soluble salt than those of the (B) zone. Above this zone, walls are sound (D). Ruiz-Agudo et al. [6] measured moisture and salt distribution inside the San Jeronimo Monastery (Granada, Spain) at two different periods, September 2004 and October 2005. They found that, at the same location, salt distribution is quite different between the two periods. In September 2004, highly soluble salts were located in the wall upper area (~200 cm height) and less soluble salts in the wall lower zone (<150 cm). At this period, there was a salt concentration gradient between the upper (lower amount) and lower (higher amount) parts of the wall. In October 2005, the total salt amount distribution was reversed, higher in the upper part than in the lower area. At this period, least soluble salts were uniformly distributed and some very soluble salts presented a maximum concentration in the middle height of the wall (100–150 cm).

Concerning salt composition, many different combinations of different salts can be found in historic buildings [7]. Sometimes, only few anions and cations are present, but in other cases a high number of chemical elements can be found in the same wall. Most common anions in buildings are sulfate, chloride, nitrate, and cations sodium, calcium, magnesium, but a great number of other cations and anions have been reported. To increase complexity, depending on the environmental conditions, different hydrated phase with the same chemical composition can be found.

All previous factors cause different decay forms in building stones that can show different development degrees. Some attempts have been done to classify these weathering forms [8,9] and to a establish damage index to quantify their intensity and extension [8].

The term "building climatology", as introduced in the seventies, deals with concepts as comfort, indoor environments, energy-saving, and sustainable building techniques. Heritage climatology deals with the long term weathering of monuments. P. Brimblecombe in 2010 [10] set down the basis of Heritage climatology, based on work that was done in the Noah's Ark project.

Knowing the behavior of building materials, especially stones, in future environments may have a great important for curators and conservators in order to select the best restauration treatments in future climate conditions. It is important to point out that this evolution will not be the same for "similar" historical buildings, depending on their geographical position (latitude and longitude, altitude, distance to the sea, etc.). In this paper, we treat only the weathering of the outside part of monuments because transfer function between outdoor meteorological data and indoor climate, will increase the complexity of the problem.

We consider the differences between past and future weathering in metropolitan France area. This evolution does not take into account particular buildings, sites, or materials. We compare the salt weathering in different locations and at different periods (past and future) without any consideration of lithology, orientation, or any other particular aspect of the stone nor the building. These locations have just been selected to represent the different climate conditions existing in France, with a spatial distribution as regular as possible. It is well known that systematic sampling is the more efficient method to obtain statistically representative results. As meteorological stations are not uniformly distributed in space, we used a nearly systematic sampling procedure while taking into account other aspects as altitude and proximity to the sea.

Salt weathering is a very complex problem that does not depend only on the climate condition, but also on the characteristics of the stone (its mineralogy, cementation degree, grain size, microstructure), on its pore system (size, connectivity), on the possible sources of salts (underground water, surrounding mortars, atmospheric pollution), and on the position of the stone in the building (orientation of the façade, height) [11]. Other important factors are local meteorological conditions as insolation or wind directions. When studying a specific monument, all of these factors should be taken into account. In our case, the goal of the study is just to obtain general tendencies.

In order to predict the evolution of salt weathering with climate change several methods have been proposed. We compared results that were obtained when applying some of these different estimation methods. Salt weathering has been estimated for single salt and salt mixture for past climate data and also for future "data" obtained by the MeteoFrance model Aladin under 4.5 scenario. The procedure, detailed in the next chapter, was applied to 41 locations in France, representative of the different climates of France.

To quantify salt weathering degradation some simplifications need to be done. Some authors use only halite salt to quantify salt weathering evolution. This is the case in the Noah's Ark project. The reason of this choice is that NaCl is a very common salt, which is present in more of weathering material and also that has only a single stable phase, which makes calculations very simple. Due to its high solubility, sodium chloride is a good salt to test salt migration, but, as shown by experimental tests, it does not produce important damage to the stone but efflorescences at the surface. It is well known that NaCl has a deliquescence relative humidity of 75.3% at 25 °C and that it does not vary very much with temperature. To quantify damage, the number of times NaCl can precipitate by a decrease of the relative humidity, is counted [12].

Another salt that was employed in the estimation of salt weathering is sodium sulfate, which is considered as one of the most dangerous salt for buildings stones. Sodium sulfate has a more complex phase diagram that sodium chloride as it can present two different stable phases at different environmental conditions, mirabilite (decahydrated) and thenardite (anhydrous) but also some metastable phases. Traditionally, it has been considered that sodium sulfate generates damage only during the transition thenardite/mirabilite but recent laboratory studies of Yu and Oguchi [13] showed that thenardite precipitation could also induce damage into the samples without any mirabilite generation during the test. In climate change studies, damage generated by sodium sulfate is estimated by the number of times thenardite is transformed into mirabilite as a consequence of temperature and relative humidity changes.

Calcium sulfate is found in most of the weathered materials in monumental stone and its weathering effect is generally recognized [14]. The kinetic of formation of this salt is very slow and it is difficult to propose estimations based on its phase diagram. Only few laboratory studies have been done to understand the weathering mechanisms of gypsum: it has been shown that under high relative humidity conditions, sulfur in the air can react very fast with limestones to form gypsum crystals in the surface of the samples [15]. Janvier-Badosa et al., 2015 [16] succeeded to form gypsum inside the samples in laboratory tests by injecting SO_2 gas in a dry sample and then allowing for water to get into the sample by capillarity. Both procedures need a high relative humidity to allow the reaction between calcite and SO_2 to precipitate gypsum. Menéndez and David [17] showed that, in wetting/drying tests with a $CaSO_4$ solution, damage is only registered under high relative humidity conditions. In previous studies, gypsum damage has been estimated by the number of days with relative humidity higher than 80%.

In natural and building environments, a combination of different salts is responsible of the weathering of stone more than a single salt. Several authors showed that wetting/drying tests that were performed with a combination of salts produce less damage than tests performed with single salt brines. Menéndez and Petráňová (2016) [18] showed using a combination of NaCl, Na_2SO_4 and $CaSO_4$ that Na_2SO_4 is much more "efficient" than a salt mixture containing the same amount of Na_2SO_4 but with NaCl and $CaSO_4$. Lindström et al. (2016) [19] have performed wetting/drying tests with Na_2SO4, $MgSO_4$ and a mixture of both and they arrive to the same conclusion, single salts generate much more damage than the combination of them. Lopez-Arce et al. (2008) [20] found that during relative humidity cycles (30–95% and 30–50%) samples that were contaminated by $MgSO_4$ suffer less damage than samples contaminated by a salt mixture of NaCl, $MgSO_4$, and $CaSO_4$. This difference is mainly due to the weathering mechanisms: formation of a thick crust with $MgSO_4$ and a fine and powdered crust that has been peeled during the cycles, with the mixture. Establishing a single estimator of weathering for complex salt mixture is not simple. To estimate salt mixture weathering as a function of environmental parameters, Menéndez (2017) [21], developed a methodology based in salt volume changes that were produced by temperature and relative humidity variations using the thermodynamic model ECOS-RUNSALT [22,23] for salt crystallization. Climate for culture project use a combination of halite and mirabilite to establish the risk of salt weathering in indoor environments [24].

2. Materials and Methods

Five different ways of quantifying salt weathering in future climate have been applied to 41 locations in Continental Metropolitan France. The localization of the selected places as well as their altitude can be seen in Figure 1a. Three of the quantification methods deal with single salts, sodium chloride, sodium sulfate, and calcium sulfate; and, two with a combination of salts, one for a

combination of the three previously cited salts and the second one for a combination of three anions
(Cl^-, SO_4^{2-}, and NO_3^-) and 4 cations (Na^+, K^+, Mg^{2+} and Ca^{2+}).

In order to obtain the future climate data, the employed methodology is quite similar to that
used in previous research [12]. A simple description of the procedure is shown here but more detailed
information can be found in [12]. Data from climate models, in our case, the Aladin model from
MétéoFrance (France), have been generated for past and future periods. In this study, we used the
temperature and relative humidity considered as the most important environmental parameters in
salt crystallization. Model "past" simulations have been compared to real data of the same period at
the same location in order to establish a downscaling protocol while taking into account specificities
of each local area. A function allowing to fit past model data to real one is established. For future
data, we have selected a RCP 4.5 scenario, which is an intermediate solution between optimistic and
pessimistic possibilities. For each location, the downscaling function obtained for past data have been
applied to future simulations. In our case the past period corresponds to 1971–2000 and future data go
from 2071 to 2100. Same scenario has been used in the climate for culture project to establish future
salt weathering risk for indoor environments [24]. Our results concern only outdoor environment.

For each location, we have calculated the mean, minimum, maximum, and standard deviation of
the daily temperature and relative humidity over 30 years. For some meteorological stations, the real
data is not available for the whole 30 years period (some years were missed). When choosing the
locations, we looked for meteorological stations with as much data as possible. With past real data and
future model data, we performed a k-mean clustering of the set of data (Tmean, Tmax, Tmin, and Tsd;
RHmean, RHmax, RHmin, and RHsd) for the 41 locations (past and future data). We used the XLSTAT
program, which runs under Excel. We gave the same weight to all variables with the determinant
criterion that allows for avoiding the scale effect for the different variables. In this way, locations have
been classified in five different classes (Figure 1b,c). The characteristics of the barycenter of each class
is presented in Table 1.

Table 1. Temperature and Relative Humidity values of the barycenter of the classes represented in
Figure 1.

Classe	Tmean	Tsd	Tmin	Tmax	RHmean	RHsd	RHmin	RHmax
1	11,101	5532	4096	19,067	81,000	5927	72,950	89,006
2	11,550	4455	6064	18,080	84,020	2656	79,914	87,913
3	13,118	6156	5693	22,680	78,117	8368	62,111	87,934
4	13,486	6426	5444	23,026	69,031	5492	59,152	75,913
5	14,383	5262	7883	22,232	76,860	2964	72,030	81,035

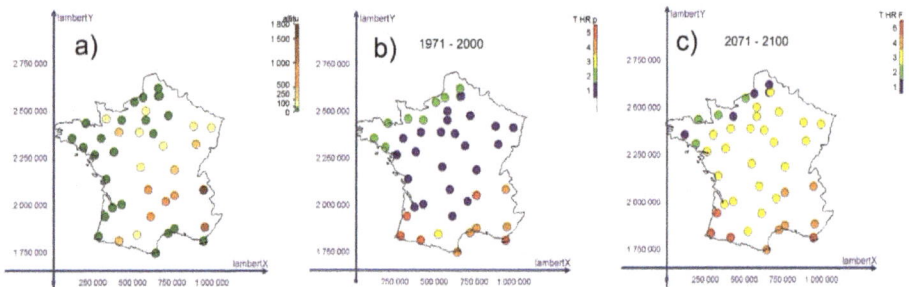

Figure 1. Position of the studied locations. (**a**) Altitude of the different points. (**b**) Classification as a
function of the temperature and relative humidity for the past and (**c**) idem for future climate.

Looking at Figure 1 we can observe that most of the locations are not very high, only two points
are located at more than 1000 m high and most of them are under 100 m. If we consider climate change,

the clearest variation is located in the central region with points that change from class 1 to class 3, with an increase of mean temperature and a decrease of relative humidity.

Next, we will explain how salt weathering has been calculated. For NaCl we just consider the number of times relative humidity change in two consecutive days between higher than 73.3% and lower than this deliquescence relative humidity. To do the calculations in all locations and for all the salts and salt combinations we used a scilab program that reads the data, performs the calculations and writes a file with the results. Even if we did not use it in this paper, the seasonality of the transitions has also be calculated (distribution of the different parameters from January to December).

For Na_2SO_4, we consider the number of times that thenardite (anhydrous phase) will convert into mirability (decahydrated phase) [25]. This occurs when environmental conditions cross the line with arrows in Figure 2. This line can be fitted by a straight line of equation RH (%) = 0.72 × T (°C) + 60.4. More details about the method can be found in [26].

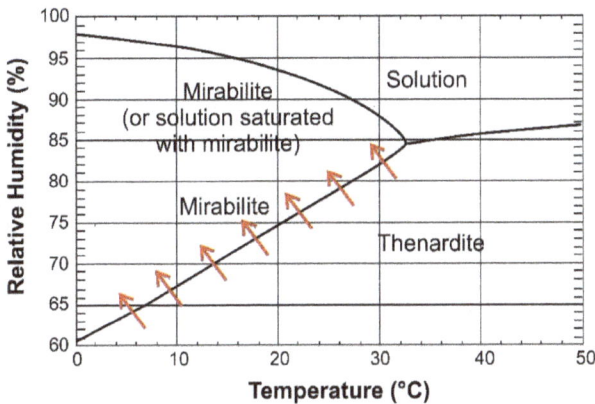

Figure 2. Phase diagram of sodium sulfate. Red arrows indicate the environmental conditions for thenardite/mirabilite transformation.

To quantify the damage produced by calcium sulfate, gypsum, we just counted the number of days relative humidity is higher than 80%. The choice of this threshold has been done based on bibliographic review, even if no many experimental studies exist about the environmental conditions that control gypsum formation in stones.

The last method used to quantify salt damage concerns the volume evolution during crystallization and phase transitions produced by environmental changes. These changes have been calculated by the ECOS-RUNSALT model and a detailed description of the procedure can be found in [21]. This method is a little bit more complex than the previous ones, but it can take into account complex salt composition. In our case, we considered a salt mixture containing a mixture of three anions (Cl^-, $SO_4{}^{2-}$, and $NO_3{}^-$) and four cations (Na^+, K^+, Mg^{2+}, and Ca^{2+}). It extracts the different environmental conditions corresponding to a volume increase of the generated salts from a standard solution. Figure 3 shows the volume of salt as a function of temperature and relative humidity. We can observe that most of damage will be produced when temperature and relative humidity decrease, mainly at low temperature and relative humidity.

Once all of the "damaging conditions" have been selected, we will compare the results to past and future data for the selected locations. The same clustering procedure that was previously used for temperature and relative humidity data has been followed for salts damage. In this way, for single salt, we consider the mean, max, min, and standard deviation of the salts transition (NaCl and Na_2SO_4) or the number of days with relative humidity higher than 80% ($CaSO_4$) for all of the locations and for both real past and model future data. With all these data, we proceed to a k-mean cluster that will classify each location in one of five classes. For the three salts procedure, we consider the same data than for single salts but the k-mean cluster procedure used the data from the three salts. In the quantification via the salt volume, we considered the number of times that volume increases between two consecutive days (mean, max, min, and standard deviation) in the k-mean clustering procedure.

Salt volume

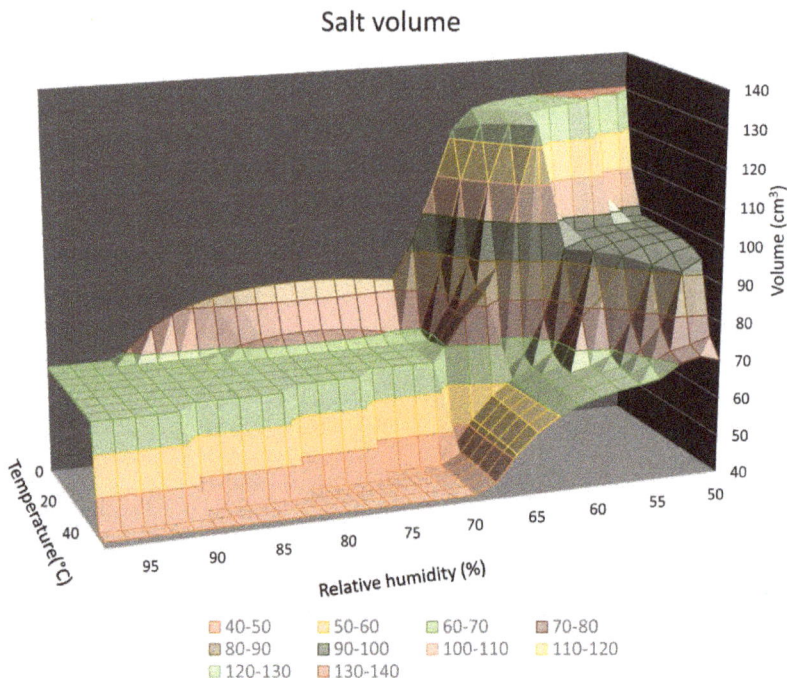

Figure 3. Salt volume as a function of temperature and relative humidity for the selected salt mixture composition.

3. Results

The main goal of this study is to compare the salt weathering previsions that were obtained by different ways of estimation. Next, we will show the results that were obtained for each procedure and we will compare them.

3.1. Salt Weathering Prevision Using Sodium Chloride

Figure 4 shows the differences in NaCl weathering in the selected locations between past (1971–2000) and future (2071–2100). Table 2 gives the statistics of the number of transitions for the barycenter of each class obtained for sodium chloride weathering.

If we consider only halite weathering, most of the locations will rest in the same class (22) or show higher salt weathering in future as compared to past period (17), only two locations show lower susceptibility to NaCl in the future than in the past. The increase in salt weathering follows a S–E to N–W direction. The number of points with the lower number of transitions (class 1) decreases from 12 in past to five in future; four of them become class 2 and three become class 3. The number of points of class 5 with the maximum of transitions, increases from 2 to 7 points.

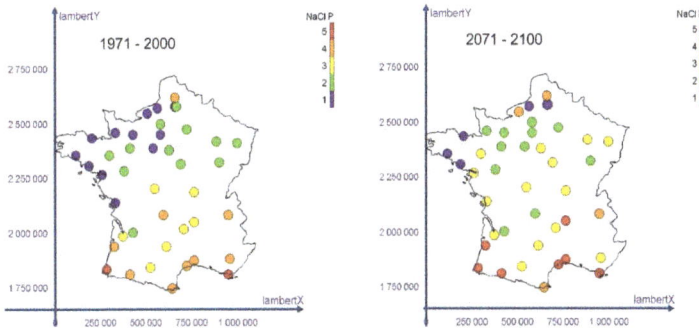

Figure 4. NaCl salt weathering estimation for past and future data.

Table 2. Number of transitions of NaCl statistical values for each class of Figure 4.

Class	Mean	Standard Deviation	Min	Max
1	60	30	17	106
2	76	47	13	148
3	91	39	33	144
4	102	26	60	144
5	119	15	93	142

3.2. Salt Weathering Prevision Using Sodium Sulfate

For sodium sulfate results are presented in Figure 5. In this case we observe very little variation, with in general a decrease of salt weathering by sodium sulfate with time. Only in 3 places in the center NE of France the salt risk increases (from green to yellow), in most of the locations it remains constant or decreases. In 26 locations, the class does not change, in 12 it decreases of one level, and only in four it increases of one level. Table 3 shows the statistics of the number of Na_2SO_4 transitions for each class.

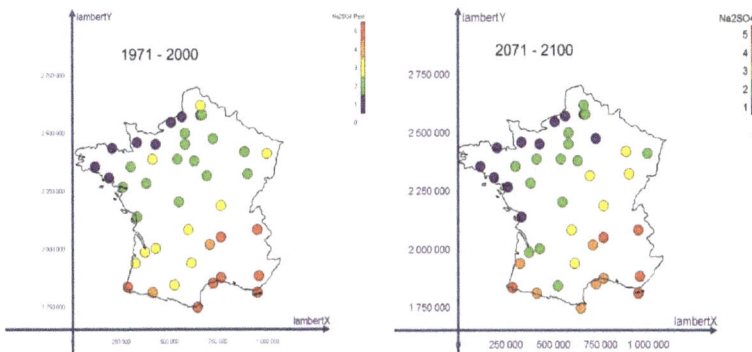

Figure 5. Na_2SO_4 salt weathering estimation for past and future data.

Table 3. Number of transitions of Na_2SO_4 statistical values for each class of Figure 5.

Class	Mean	Standard Deviation	Min	Max
1	31	26	2	76
2	50	40	3	111
3	67	49	5	140
4	78	27	34	113
5	100	24	64	136

3.3. Salt Weathering Prevision Using Calcium Sulfate

For calcium sulfate, we do not use the number of transitions, but the number of days with high relative humidity, which can allow the penetration of water into the pore system and then the precipitation of gypsum inside the stones. The results in Figure 6 show more variation between the past and the future than with NaCl and Na$_2$SO$_4$. We can observe a decrease of the risk in the central part of France, which is mainly due to the reduction of the relative humidity in these regions, from a mean of 78% to 69%, as shown of Figure 1 and Table 1. Concerning the spatial distribution of risk, it is opposite to the previous two cases, with more vulnerable regions in the NW coast and the least ones in the Mediterranean coast. In this case, 18 locations do not change of risk class, 21 decrease in one level, two in two levels, and only in one location the risk due to gypsum increases of one level. Table 4 gives the statistics of the number of the number of days with RH>80% for the barycenter of each class obtained for sodium sulfate weathering.

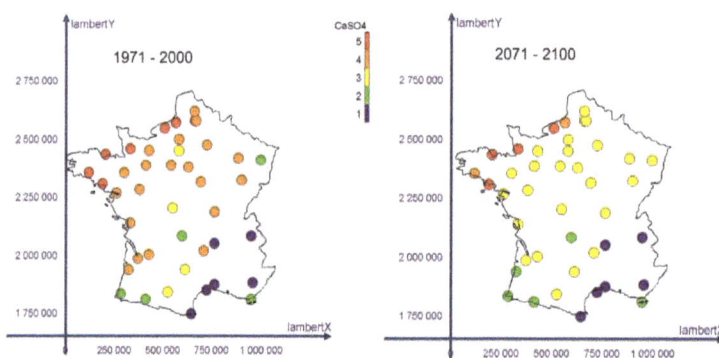

Figure 6. CaSO$_4$ salt weathering estimation for past and future data.

Table 4. Number of days with RH higher than 80% used to estimate CaSO$_4$ weathering. Statistical values for each class of Figure 6.

Class	Mean	Standard Deviation	Min	Max
1	219	100	71	383
2	388	95	231	529
3	446	228	69	759
4	497	173	267	765
5	636	94	483	777

3.4. Salt Weathering Prevision Using a Combination of Sodium Chloride, Sodium Sulfate and Calcium Sulfate

If we apply the clustering procedure to a combination of previous three salts data, we obtain different results. We can say that in the NW (lowest risk), and SW coasts and in the Mediterranean (highest risk) area, between past and future, weathering risk stays constant or decreases but in central France the risk mainly increases. In general we can say that in 19 places salt weathering risk does not change, in four locations it decreases and in 19 locations it increases.

Looking at Table 5, we can see how NaCl and Na$_2$SO$_4$ number of transitions increases from least risk zones to highest one but CaSO$_4$ risk decreases.

Table 5. Number of NaCl and Na$_2$SO$_4$ transitions and number of days with RH > 80% (CaSO$_4$) for each class of Figure 7.

Classe	NaCl Mean	Na$_2$SO$_4$ Mean	CaSO$_4$ Mean
1	64	28	636
2	76	50	448
3	79	51	512
4	103	73	418
5	105	97	238

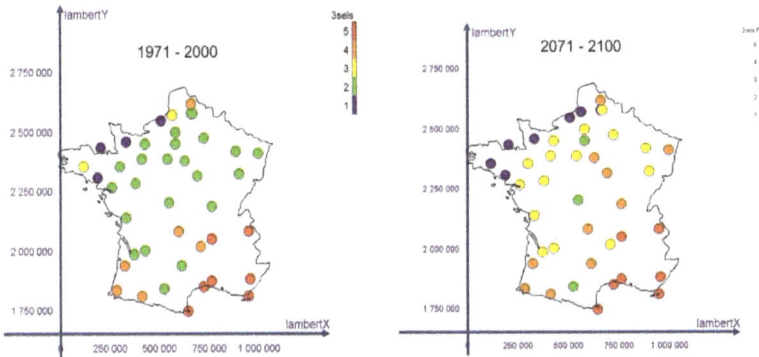

Figure 7. Salt weathering estimation for past and future data taking into account NaCl, Na$_2$SO$_4$, and CaSO$_4$ at the same time.

3.5. Salt Weathering Prediction Using the Volume Increase of Precipitated Salts

In this case, we use several different salts, as can be seen in [21]. The involved salts are sodium, magnesium, and calcium sulfates, magnesium, sodium, and potassium nitrates, sodium chloride, a double salt of sodium sulfate and nitrate and another of sodium and magnesium sulfate. Results are shown in Figure 8. In general salt risk will increase with time. At the end of 21th century as compared to end of 20th century, in 14 locations salt weathering risk remains constant and in the other 24 it increases. These increases are higher than in the previous cases. It increases of one level in 13 locations, of two levels in eight locations, of three levels in six locations and in one location it increases of four levels. The results show that salt weathering risk is lower in all the west coast than in the rest of France. SE part of France presents the most vulnerable area.

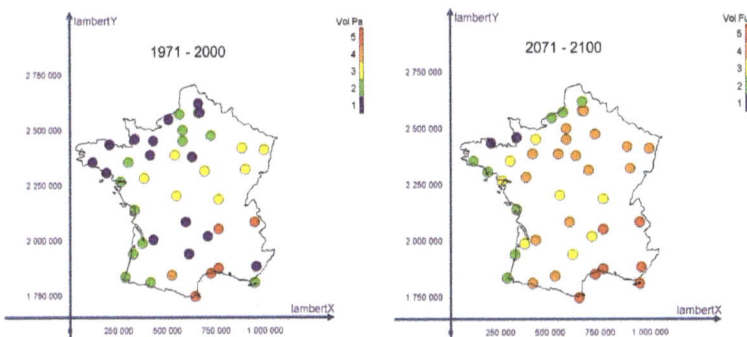

Figure 8. Salt weathering estimation for past and future data taking into account the volume change of salts during environmental conditions changes.

4. Discussion

We have shown that when estimating salt weathering from climate data, the use of different methods may conduct to different results. For example, if we consider the estimations of NaCl or Na$_2$SO$_4$ salt weathering, the spatial distribution for past period is quite similar with three SW–NE bands dividing the country and with a NW to SE gradient; when using CaSO$_4$, the gradient has the same orientation but in the opposite direction. If we look at the evolution of salt weathering with time, for different salts the evolution is not the same. For NaCl the general trend is a slight increase of the weathering but for Na$_2$SO$_4$ the tendency is not so clear. We found a diminution of salt weathering in the West coast but weathering remains constant in the S–E, with maybe a slight decrease of the number of transitions. In a small central-East part, future weathering may be more important than in the past. These results are not in agreement with previous works [12] for two cities in the NW coast of France. Previous research found that halite weathering may remain constant during the 21th century, but the number of thenardite/mirabilte transitions will increase. In our case we found that the total number of transition increases for sodium chloride and decreases for sodium sulfate.

If we look at the results for calcium sulfate, the spatial distribution of damage is also organized in three areas similar to the previous one, with a gradient from South–East to North–West. We consider that zones with high relative humidity are more propitious to gypsum generation than dryer ones. The main change with future climate is a decrease of gypsum weathering risk in the central regions. N–W and S coast do not present major changes.

In the third clustering process, we took into account data of the three salts (NaCl, Na$_2$SO$_4$, and CaSO$_4$). The obtained results also show a geographical distribution with a NW to SE gradient. In this case, we observe slight differences between past and future in the N–W coast, not evident for individual salts. The most important changes take place in the central region, going from class 2 to 3 or 4. As in previous cases, southern locations do not show changes in salt weathering in the future.

The last estimation of salt weathering, based on the volume change, gives different results from the previous ones. The N–W coast, as in most of the previous cases, presents the lowest weathering susceptibility. The most important weathering takes place in the Mediterranean coast and locations with high altitude. The distribution in the central part of France is not as regular as before. Concerning changes in future weathering, a general increase of weathering is observed between the end of XX century and the end of XXI century. We can consider that this estimator is the one with predicting the highest changes between past and future.

5. Conclusions

We have shown in this paper that different estimators that are used to quantify salt weathering can lead to significantly different results. Results that were obtained for gypsum are quite different from those obtained for halite or thenardite/mirabilite. The gypsum weathering mainly depends on relative humidity, as it will precipitate inside the stone, causing damage when RH is high. This result is in accord with recent research showing that gypsum will precipitate inside the pores when SO$_2$ from the atmosphere reacts with water inside the pores. This water can come from capillarity suction or by condensation due to an increase in the RH. In our estimations, we did not take into account rain in a direct way but via relative humidity. We can consider that, in rainy days, relative humidity will increase.

Experimental studies [18] have shown that weathering by sodium chloride and sodium sulfate is more important when relative humidity decreases under the 80% threshold. When comparing Figure 1 to Figures 4–6, it is clear that salt weathering distributions is highly dependent on temperature and relative humidity.

Figure 9 shows the increase (positive) or decrease (negative) of salt weathering between the future and the past. "Predicting" salt weathering change in future, basing the study in only one salt, may conduct to over or more probably, under estimations. We consider that taking into account as much salts as possible is the better way to quantify general salt weathering, because most of the time.

In cultural heritage buildings, we do not find a single salt but a combination of them. Differences observed between the results that were obtained with three salts and volume changes may be explained by the fact that in the volume calculations we took into account other salt compositions with nitrates and magnesium and potassium. As can be seen in Figure 10, at 25 °C, and if we do not consider gypsum, the maximum volume increase takes place when relative humidity decreases between 70% and 55%. This fact explains why maximum salt damage takes place in dryer climates. Climate for culture project found [24] a general decrease of salt weathering for most of France during the same period.

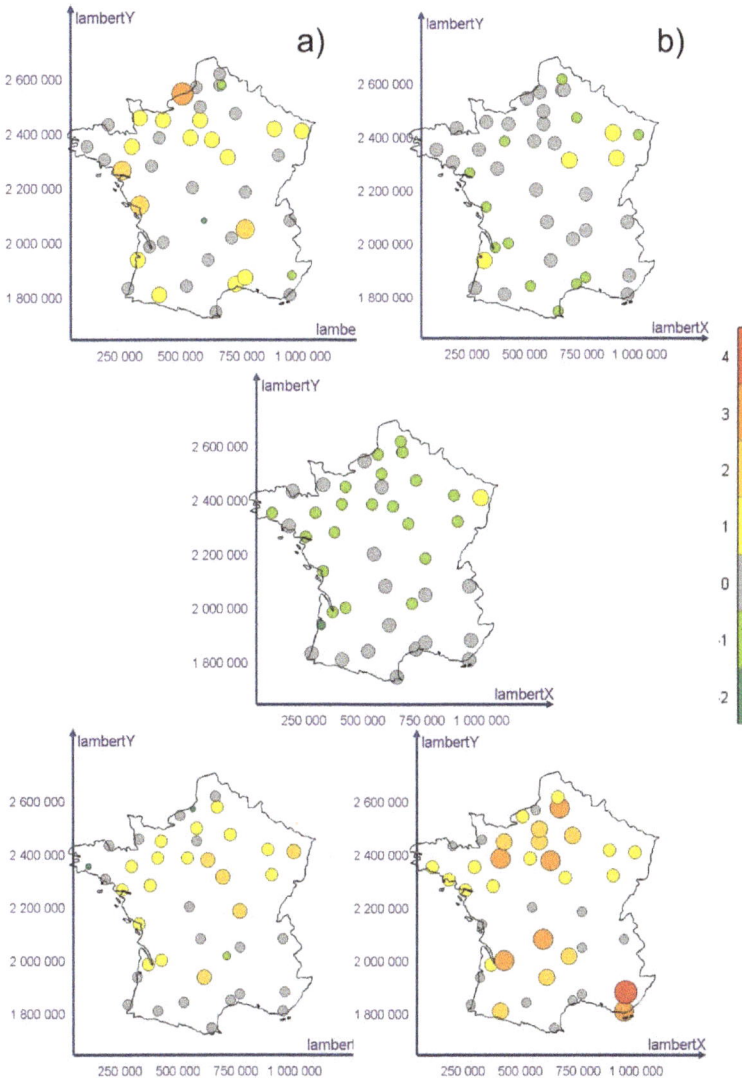

Figure 9. Salt weathering variation between past and future data taking into account: (**a**) NaCl, (**b**) Na_2SO_4, (**c**) $CaSO_4$, (**d**) Three previous salts, and (**e**) global volume change.

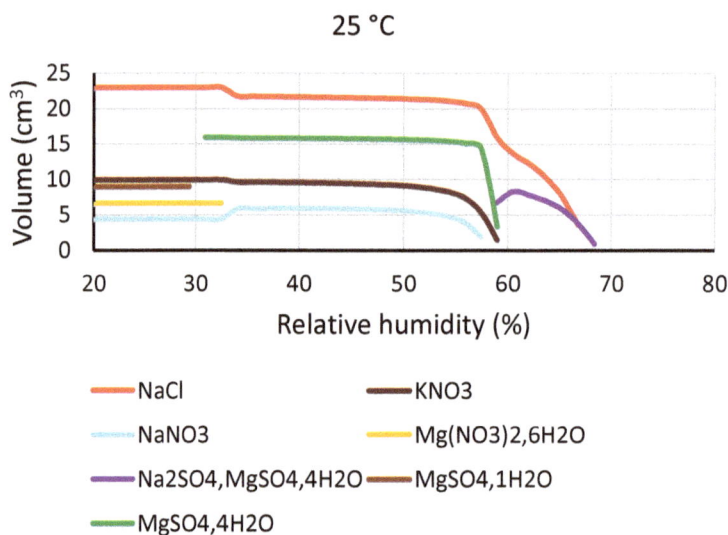

Figure 10. Salt volume variation for the selected composition without gypsum, calculated with ECO-RUNSALT model.

It is important to remember that the obtained results do not concern specific buildings nor materials. The main interest of the presented results consists in determining how weathering can change from one region to another and from past time to future, if we consider a same material being exposed to the same exposition, with the same pollutant concentration, etc. In real case studies, many other parameters need to be considered as the nature of the stone and the surrounding materials, nature and concentration of salts, actual and past pollution, orientation of the façade, underground water nature and conditions, and many others.

Computational facilities can help when doing the estimations. A small program, in scilab in our case, can considerably reduce the calculation time of transitions number and avoid the use of other approximations based on mean temperature or relative humidity, as proposed before [12].

Funding: This research received no external funding.

Acknowledgments: We thank Météo-France for the provided data especially Michel Déqué for the climate model realizations. We thank also Christian David (University of Cergy-Pontoise) for his help in writing the article.

Conflicts of Interest: The authors declare no conflict of interest.

References

1. Sallam, E.S.; El-Aal, A.K.; Fedorov, Y.A.; Bobrysheva, O.R.; Ruban, D.A. Geological heritage as a new kind of natural resource in the Siwa Oasis, Egypt: The first assessment, comparison to the Russian South, and sustainable development issues. *J. Afr. Earth Sci.* **2018**, *144*, 151–160. [CrossRef]
2. Steiger, M. Salts and crusts. In *The Effects of Air Pollution on the Built Environment*; Imperial College Press: London, UK, 2003; pp. 133–181.
3. Charola, A.E.; Ware, R. Acid deposition and the deterioration of stone: A brief review of a board topic. In *Natural Stone, Weathering Phenomena, Conservation Strategies and Case Studies*; Special Publication; Geological Society: London, UK, 2002; pp. 393–406.
4. Charola, A.E. Salts in the deterioration of porous materials: An overview. *J. Am. Inst. Conserv.* **2000**, *39*, 327–343. [CrossRef]
5. Arnold, A.; Zehnder, K. Monitoring wall paintings affected by soluble salts. In *The Conservation of Wall Paintings*; Getty Conservation Institute: Los Angeles, CA, USA, 1991; pp. 103–135, ISBN 0-89236-162-X.

6.   Ruiz-Agudo, E.; Lubelli, B.; Sawdy, A.; van Hees, R.; Price, C.; Rodriguez-Navarro, C. An integrated methodology for salt damage assessment and remediation: The case of San Jeronimo Monastery (Granada, Spain). *Environ. Earth Sci.* **2011**, *63*, 1475–1486. [CrossRef]

7.   Goudie, A.; Viles, H.A. Nature of salt involved in salt weathering and sources of moisture. In *Salt Weathering Hazards*; Wiley: Hoboken, NJ, USA, 1997; 256p, ISBN 978-0-471-95842-0.

8.   Fitzner, B.; Heinrichs, K. Damage diagnosis on stone monuments—Weathering forms, damage categories and damage index. In *Abstracts of the International Conference "Stone Weathering and Atmospheric Pollution Network (SWAPNET)"*; Prikryl, R., Viles, H.A., Eds.; Prachov Rocks: Jicin, Czech Republic, 2001.

9.   ICOMOS-ISCS. *Illustrated Glossary on Stone Deterioration Patterns*; Belmin, V., Ed.; ICOMOS: Paris, France, 2008; ISBN 978-2-918086-00-0.

10.   Brimblecombe, P. Heritage climatology. In *Climatechange and Cultural Heritage*; Lefevre, R.A., Sabbioni, C., Eds.; Edipuglia: Bari, Italy, 2010; pp. 57–64.

11.   Sanjurjo-Sanchez, J.; Alves, C. Decay effects of pollutants on stony materials in the built environment. *Environ. Chem. Lett.* **2012**, *10*, 131–143. [CrossRef]

12.   Grossi, C.M.; Brimblecombe, P.; Menéndez, B.; Benavente, D.; Harris, I.; Deque, M. Climatology of salt transitions and implications for stone weathering. *Sci. Total Environ.* **2011**, *409*, 2577–2585. [CrossRef] [PubMed]

13.   Yu, S.; Oguchi, C.T. Is sheer thenardite attack impotent compared with cyclic conversion of thenardite–mirabilite mechanism in laboratory simulation tests? *Eng. Geol.* **2013**, *152*, 148–154. [CrossRef]

14.   Charola, A.E.; Puehringer, J.; Steiger, M. Gypsum: A review of its role in the deterioration of building materials. *Environ. Geol.* **2007**, *52*, 339–352. [CrossRef]

15.   Gibeaux, S.; Thomachot-Schneider, C.; Eyssautier-Chuine, S.; Marin, B.; Vazquez, P. Simulation of acid weathering on natural and artificial building stones according to the current atmospheric SO_2/NOxrate. *Environ. Earth Sci.* **2018**, *77*, 327. [CrossRef]

16.   Janvier-Badosa, S.; Beck, K.; Brunetaud, X.; Guirimand-Dufour, A.; Al-Mukhtar, M. Gypsum and spalling decay mechanism of tuffeau limestone. *Environ. Earth Sci.* **2015**, *74*, 2209–2221. [CrossRef]

17.   Menéndez, B.; David, C. The influence of environmental conditions on weathering of porous rocks by gypsum: A non-destructive study using acoustic emissions. *Environ. Earth Sci.* **2013**, *68*, 1691–1706. [CrossRef]

18.   Menéndez, B.; Petráňová, V. Effect of mixed vs. single brine composition on salt weathering in porous carbonate building stones for different environmental conditions. *Eng. Geol.* **2016**, *210*, 124–139. [CrossRef]

19.   Lindström, N.; Heitmann, N.; Linnow, K.; Steiger, M. Crystallization behaviour of $NaNO_3$—Na_2SO_4 salt mixtures in sandstone and comparison to single salt behavior. *Appl. Geochem.* **2015**, *63*, 116–132. [CrossRef]

20.   Lopez-Arce, P.; Doehne, E.; Martin, W.; Pinchin, S. Magnesium sulfate salts and historic building materials: Experimental simulation of limestone flaking by relative humidity cycling and crystallization of salts. *Mater. Constr.* **2008**, *58*, 289–290. [CrossRef]

21.   Menéndez, B. Estimation of salt mixture damage on built cultural heritage from environmental conditions using ECOS-RUNSALT model. *J. Cult. Herit.* **2017**, *24*, 22–30. [CrossRef]

22.   Price, C. *An Expert Chemical Model for Determining the Environmental Conditions Needed to Prevent Salt Damage in Porous Materials*; Archetype: London, UK, 2000; p. 136, ISBN 1873132522.

23.   Bionda, D. Runsalt Computer Program. 2002–2005. Available online: http://science.sdf-eu.org/runsalt/ (accessed on 2 November 2018).

24.   Leissner, J.; Kilian, R.; Kotova, L.; Jacob, D.; Mikolajewicz, U.; Broström, T.; Ashley-Smith, J.; Schellen, H.L.; Martens, M.; van Schijndel, J.; et al. Climate for Culture: Assessing the impact of climate change on the future indoor climate in historic buildings using simulations. *Herit. Sci.* **2015**, *3*, 38. [CrossRef]

25.   Benavente, D.; Brimblecombe, P.; Grossi, C.M. Salt weathering and climate change. In *New Trends in Analytical, Environmental and Cultural Heritage Chemistry, Transworld Research Network*; Colombini: London, UK, 2008; Chapter 10; pp. 277–286.

26.   Menendez, B. Salt climatology applied to built Cultural Heritage. In *Cultural Heritage from Climate Change to Global Change*; Lefevre, R.A., Sabbioni, C., Eds.; Edipuglia: Bari, Italy, 2016; pp. 35–50.

geoscience

MDPI

Article

δ^{13}C and δ^{18}O Stable Isotope Analysis Applied to Detect Technological Variations and Weathering Processes of Ancient Lime and Hydraulic Mortars

Elissavet Dotsika [1,2,*], Dafni Kyropoulou [1,3], Vassilios Christaras [1,3] and Georgios Diamantopoulos [1]

[1] Institute of Nanosciences and Nanotechnology, Stable Isotope Unit, National Centre for Scientific Research, Demokritos, Agia Paraskevi, 15310 Attiki, Greece; d.kyropoulou@inn.demokritos.gr (D.K.); christar@auth.gr (V.C.); g.diamantopoulos@inn.demokritos.gr (G.D.)
[2] Institute of Geosciences and Earth Resources, Via G. Moruzzi 1, 56124 Pisa, Italy
[3] School of Geology, Faculty of Sciences, Aristotle University of Thessaloniki, GR-541 Thessaloniki, Greece
* Correspondence: e.dotsika@inn.demokritos.gr

Received: 18 July 2018; Accepted: 4 September 2018; Published: 8 September 2018

Abstract: Samples of mortars were collected from lime and hydraulic mortars affected by environmental degradation. A total of 63 samples were obtained from Hellenistic, Late Roman and Byzantine historic constructions located at Kavala, Drama and Makrygialos in North Greece. Samples were collected in sections from the surface up to 6 cm deep using a drill-core material. The first sample was collected from the external layer, while the internal samples were collected each 1 cm beeper from the previous, in order to monitor the moisture ingress. Isotopic data will make it possible to create an ideal Hellenistic and Byzantine mortar layer and to provide weathering gradients. The isotopic values comprise a range of δ^{13}C and δ^{18}O values from $-17.1‰$ to $1.2‰$ and $-25.9‰$ to $-2‰$, respectively. The weathering process of Hellenistic and Byzantine are expressed, by the regression lines δ^{18}O$_{\text{calcite matrix}} = 0.6 \times \delta^{13}C_{\text{calcite matrix}} - 1.9$ and δ^{18}O$_{\text{calcite matrix}} = 0.6 \times \delta^{13}C_{\text{calcite matrix}} - 2.0$ for hydraulic and Lime mortars respectively. Pronounced isotopic shift to heavy or light δ^{13}C and δ^{18}O in the carbonate matrix was attributed to the primary source of CO_2 (atmospheric versus biogenic) and H_2O (evaporation of local primary water), in residual limestone and in secondary processes such as recrystallization of calcite with pore water and salts attack. Exogenic processes related to biological growth are responsible for further alterations of δ^{18}O and δ^{13}C in lime mortars. This study indicated that stable isotope analysis is an excellent tool to fingerprint the origin of carbonate, the environmental setting conditions of mortar, origin of CO_2 and water during calcite formation and to determine the weathering depth and the potential secondary degradation mechanisms.

Keywords: stable isotopes; δ^{13}C; δ^{18}O; mortars; mineralogy; degradation

1. Introduction

It is very important to respond to major new challenges regarding the conservation of cultural heritage, by predicting deterioration features and evaluating the nature of damages monuments. Its preservation demonstrates recognition of the necessity of the past while on a different level ensures a mean for the validation of human-memories and confirms human history. The material nature of monuments is what makes it so important, allowing a direct interaction with human senses, yet it also entails that it is a constant state of chemical transformation, making its preservation an everlasting task of continuously growing importance. Hydraulic and lime mortars constitute main structural components in architectural constructions. Environmental degradation of historic mortars are a main threat for the preservation of historic monuments [1–6]. The main agents of decay (such as acid attack, leaching action, salts attack, damage due to frost and fire, insufficient mixing and choice of constituents) can cause extensive cracks and total disintegration of historic constructions [7–10]. There are two main categories of limes, hydraulic and non-hydraulic. Lime is formed by burning a source of calcium carbonate such as limestone to evolve carbon dioxide and form calcium oxide (1) [(CaO, quicklime)]. The calcium oxide is then slaked with water under heating to produce calcium hydroxide (slaked lime, $Ca(OH)_2$) (2). This mass will be then mixed with aggregate materials such as sand, gravel, crushed stone or iron blast-furnace slag and various organic and inorganic additives such as pozzolanic materials or mud, forming calcium carbonate (3). The mortar hardens with the absorption of carbon dioxide from the atmosphere gaining strength. The formation of calcium carbonate mortars is described in the following reactions:

$$CaCO_3 \rightarrow CaO + CO_2 \tag{1}$$

$$CaO + 2H_2O \rightarrow Ca(OH)_2 + H_2O \tag{2}$$

$$Ca(OH)_2 + CO_2 \rightarrow CaCO_3 + H_2O \tag{3}$$

The reactions within the formation of hydraulic compounds are described below:

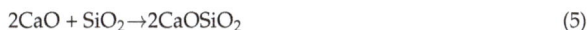

$$CaO + Al_2O_3 \rightarrow CaOAl_2O_3 \tag{4}$$

$$2CaO + SiO_2 \rightarrow 2CaOSiO_2 \tag{5}$$

The sources of mortar decay could be physical, related to physical variations of water inside masonry (evaporation, capillary flow, ice formation, etc.) or chemical (formation of expansive products such as ettringite and thaumasite).

Also, lime-based mortars can be easily attacked by sulphur compounds. Sulfation occurs when a layer of gypsum is formed in the surface of calcite that is more soluble than calcite (6).

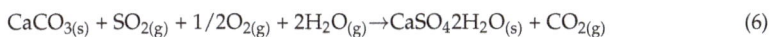

$$CaCO_{3(s)} + SO_{2(g)} + 1/2O_{2(g)} + 2H_2O_{(g)} \rightarrow CaSO_4 2H_2O_{(s)} + CO_{2(g)} \tag{6}$$

Therefore, in the presence of water (rain) this gypsum layer is dissolved, causing dissolution of the external calcite layer. Furthermore, concrete deterioration can be related to internal sulfate attack. Aggregates containing sulfide minerals (pyrite) can become an internal source of sulfate ions. These ions can promote the development of chemical reactions with the cement paste composition, which may result in the formation of expansive and deleterious products that are typical of sulfate attack known as ettringite, gypsum and thaumasite [11,12].

In order to determine the above degradation mechanisms and propose suitable conservation strategies, it is essential to examine the calcium carbonate origin and to identify the possible causes that changed the initial composition. Stable isotopic data provides information of the primary sources of raw material and offers considerable potential in the investigation of lime mortar and lime plaster (composed of sand, water, and lime and often contained horse hair for reinforcement and pozzolan additives). In particular, stable isotope fingerprinting is a concept that has been successfully developed and applied in earth sciences for several decades [13,14]. The elements C and O have a characteristic isotope ratio that varies slightly but significantly when transformations like phase changes or chemical reactions occur. Those slight variations can be measured and allow for valuable conclusions to be drawn on the origin and the history of these elements. Previous studies on historical mortars implemented mineralogical and chemical analysis, evaluation of the aggregate to matrix ratio and provenance of raw materials [15–18]. However, studies focused on stable isotope analysis of lime mortars are rare [19,20]. Furthermore, little is known about the isotopic change of the initial C and O in carbonate matrix that was caused by alteration of the primary source of CO_2 and H_2O in mortar over time. Human influence and biological growth are major exogenic processes which alter O and C in lime mortar. In particular oxygen isotope is common between CO_2 and H_2O and has the potential to elucidate the history of primary source of H_2O or alteration processes with water of meteoric origin, while carbon isotope composition can identify the origin of CO_2. Finally, the main objective of this study is to examine the sources of mortar decay in four Hellenistic, Roman, Late Roman and Byzantine masonries. The determination of the exact origin of the materials is the key to reach the above objective and this can be achieved using stable isotopes. Furthermore, stable isotope analysis will allow examining what has changed relatively to environmental conditions and secondary processes from ancient time until today, to distinguish the nature of the structural materials used (mortar) and to disambiguate the causes that incurred the degradation phenomena.

2. Methods and Materials

2.1. Methods

Mortar lumps were analyzed using Scanning Electron Microscopy (SEM) with Energy Dispersive X-ray (EDX) analysis since this technique is considered a powerful tool to characterize the microstructure of the materials [21,22].

Major and minor elements for polished mortar samples were analyzed using a FEI/Quanta Inspect D8334 SEM, FEI company, Hillsboro, OR, USA, fitted with a low vacuum chamber and an energy dispersive X-ray analyzer at NCSR "Demokritos", Athens, Greece. The operating conditions were a 25 KV accelerating potential and a 107 μA emission current, giving a count rate of 400 cps on metallic cobalt, the working distance was normally 10 mm. The detector was calibrated with a range of pure elements, synthetic oxides and well characterized minerals. The mortar samples were vacuum-dried, polished in resin blocks, mounted onto metal stubs and coated with a thin film of gold and examined with the SEM.

The elements present in the mortar samples were determined using X-ray measurements of each particle. A backscattered electron image was obtained in every sample and qualitative analysis of 10–20 individual particles was performed. For bulk analysis of mortars, the SEM beam was rastered on each sample at 100× over a 4 × 3 mm sized area. Accuracy and precision in bulk analysis for mortar samples were expected to be somewhat reduced because mortars are heterogeneous materials, however for comparative purposes the results were normalized to 100% and are considered to give an approximation of the major oxides. The imaging was performed using four-quadrant Backscattered Electron Imaging (BSE). The backscattered images enabled the determination of the structural components in mortars, since heavy elements such as iron and lead appear brighter than light elements such as carbon or oxygen. As a result, backscattered electron images demonstrate the make-up of mortars, because sharp changes in grey levels represent different compounds.

Mineralogical analysis was achieved using optical microscopy [22,23]. Optical microscopy was performed using a polarizing microscope at the Department of Mineralogy, Petrology and Economic Geology at Aristotle University of Thessaloniki. Thin sections were prepared by drying the samples (lumps) using acetone in vacuum. The samples were vacuum impregnated with epoxy resin, cut with a slow-speed diamond saw to minimize damage and then mounted on glass slides and ground to 30 mm thickness. Observations were made under crossed polars (XPL) using a Leitz Laborlux 11 POL S, (Leica Microsystems' Microscopy and Scientific Instruments, Wetzlar, Germany). The aggregate clasts, and the fine-grained calcite, may be identified using common geological techniques [24,25]. The terminology developed for the description of archaeological ceramics was applied to some extend in mortar cross section investigations [26,27].

The isotopic analyses took place at Stable Isotope Unit of Institute of Nanosciences and Nanotechnology (NCSR "Demokritos") on a Thermo Delta V Plus IRMS equipped with GasBench II device. Mortar samples were separated using scalpels in two or three sections reflecting the changes in isotopic composition from interior to exterior. Small quantities of clean (leached with acetic acid) carbonate material were separated and after grinding to a fine powder were diluted in ortho-phosphoric acid and the CO_2 produced was measured in the isotope ratio mass spectrometer. The isotopic results are reported in the usual delta terminology versus VPDB isotopic standard, delta being defined as follows: δ = [(Rsample − Rstandard)/Rstandard] × 1000 where R is the ratio between the heavy and the light isotope, in this case $^{18}O/^{16}O$ or $^{13}C/^{12}C$. The reported values are the mean of two or more consistent measurements of each sample. The standard deviation of the materials measurements is very good, ranging on average between ±0.1‰ and ±0.2‰ [28–30].

2.2. Materials

A total of 63 samples were obtained from four Hellenistic, Late Roman and Byzantine historic constructions located at Kavala, Drama and Makrygialos in North Greece (Figure 1). The location of the studied sites is considered maritime therefore all the architectural constructions and all mortar samples are seriously affected by salt crystallization [31,32]. Also, extensive fragmentation is observed in fortification walls in Drama, in Marmarion tower in Kavala and in funerary monuments in Makrygialos. Different types of crust are also observed in the surface of the mortars. Dark soiling patterns are observed on the mortars surface. Black crusts are likely to be derived from secondary recrystallization of calcitic binders while green crusts are associated with biodegradation and biological growth. The samples were collected in sections by pressing a drill-core material 6 cm towards the surface. The first sample was collected from the surface, while the internal samples were collected each 1cm beeper from the previous. The samples collected from the exterior layers are exposed to environmental degradation while the samples collected from interior layers do not possess any evidence of degradation.

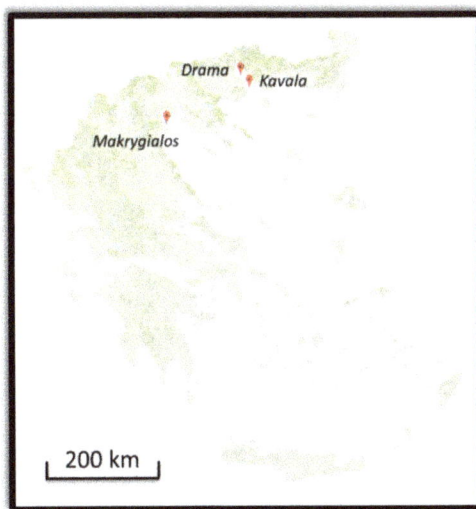

Figure 1. Map of sampling locations. Makrygialos (40°25′ N 22°36′ E), Drama (41°9′ N 24°8′ E), Kavala (40°56′ N 24°24′ E).

2.2.1. Anaktoroupoli and Land Walls in Kavala

Samples (AN1a-1 to Anp-1) were obtained from Anaktoroupoli which is an area located in the region of Nea Peramos in Kavala, Greece. In specific, Anaktoroupoli is the remaining of a Byzantine town located in the west coast of Kavala. It is dated at 14th century AD based on the characteristics of the masonry and the features observed in the coins excavated in the region [33]. The samples collected from this site were joints. The joints of the brickwork structure are 2 ± 4 cm thick, and probably have a bearing function [34]. A few samples were also collected from land walls in Kavala (KA1b, KA5 and KA6), but these were not included in stable isotope analysis, because of analytical errors (the samples

KA1b, KA5, KA6 were entirely covered with salt and the SEM detector could not detect the other constituents). Land walls in Kavala were constructed in four different historical periods; however samples selected were characterized as late Byzantine (9th to 10th century).

2.2.2. Marmarion Tower in Kavala

Samples (MA1-3 to MA5-2) were collected from Marmarion tower in Kavala, Greece. The Marmarion tower was built in 1367 by Priests Alexios and Ioannis for the Ancient Monastery of Pantokratoras. The tower is constructed using secondary ancient material, mainly sandstones, marlstones, fired bricks and good quality lime mortar. The ancient material is composed of large marlstones from architectural parts such as inscribed columns [34].

2.2.3. Fortification Walls in Drama

Samples (DRa-1 to DR2g-1) were obtained from joints located at fortification walls that surround the historical center of Drama, one of the most important monuments of the region [34]. Two main parts of the walls have survived: the northern parts have been almost entirely preserved, apart from the north-eastern corner and all of the west part. The walls were founded in Late Roman times (3th–6th century AD) and later building phases are ascribed to the Frankish period. Modern building materials were identified, added possibly within restoration treatments. This study will examine samples from the Later Roman building phase.

2.2.4. Funerary Monuments in Makrygialos

Mortar samples (MK1-1 to MK10-2) were obtained from 4th century BC Hellenistic tombs in the area of Pydna in North Greece. Ancient Pydna lies in the North of Pieria, two kilometers away from South Makrygialos. The monuments were discovered in rescue excavations under the supervision of KZ' Ephoreia of Classical and Prehistoric Antiquities [35].

3. Results

3.1. Micromorphological and Elemental Analysis

3.1.1. Anaktoroupoli-Land Walls Kavala

Backscattered electron photomicrographs of sample AN1b, representative of the group of samples obtained from Anaktoroupoli, Nea Peramos, illustrates the morphology of a Byzantine mortar. The photomicrograph (Figure 2) depicts large irregularly shaped grains acting as aggregates, while the matrix is composed from homogeneous smaller grains. The sample is composed of 50.7% quartz and 31.7% CaO according to bulk analysis of oxides in the SEM and the ternary plot with EDX values (Table 1). Petrographic analysis elucidated that the matrix is composed of microcrystalline calcite (Figure 3). Thin section demonstrates that this is a coarse mortar, the size of mineral components rises up to 6 mm. The mineralogical composition of the aggregates in this sample includes quartz, calcite, plagioclase, potassium feldspar, microcline, muscovite and traces of sericite and kaolinite. Thin section photomicrograph indicates that quartz is the most abundant mineral, in line with SEM analysis; quartz grains reach up to 3 mm. Quartz is usually coarse and shows a strong recrystallization layer on the boundaries of quartz grains. Most of the mineral components demonstrate rounded cavities surrounded by a thin layer of calcite. Backscattered electron photomicrograph reveals fibrous crystals of sericite (indicated with black arrow) with hexagonal structure and needle morphology (10 mm in length) are present as parts of the aggregate fraction. Calcitic matrix shows intensive fragmentation and cracks attributable to the degradation of mortar over time. Extensive deposition of salts is observed in sample K5, obtained from land walls in Kavala. The surface of the mortar is entirely covered with salts. Point analysis detected the elements Na and Cl on the entire surface of the sample.

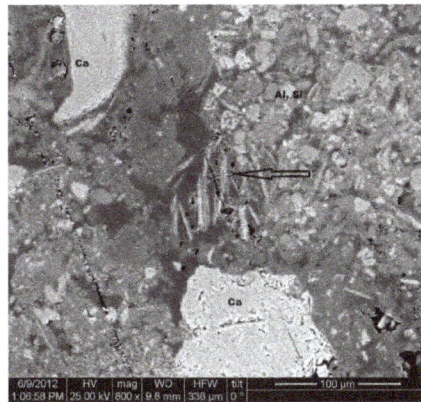

Figure 2. SEM/Backscattered Electron Imaging (BSE) (×253) photomicrograph of sample AN1b from land walls in Kavala. The black arrow indicates fibrous crystals of sericite.

Table 1. Data in weight per cent oxide normalized to 100%.

Sample	Location	Al_2O_3	SiO_2	CaO	MgO	Cl_2O	K_2O	Na_2O	BaO	Fe_2O_3
AN1b	Anaktoroupoli, Kavala	10.4	50.7	31.7	na	0.6	3.2	2.4	na	na
DR2c	Drama	9.8	57.6	13.0	13.2	1.4	na	na	na	5.1
Dr3	Drama	3.5	49.0	49.0	1.3	1.1	1.1	na	na	na
KA1b	Land walls, Kavala	10.4	50.7	31.7	na	0.6	na	2.4	na	na
KA5	Land walls, Kavala	5.3	29.8	54.5	na	2.4	na	3.8	na	na
KA6	Land walls, Kavala	7.6	38.7	na	0.5	22.2	3.3	26.3	1.5	na
MA2	Marmarion Tower, Kavala	9.8	52.9	33.4	na	na	na	na	na	na
MA6	Marmarion Tower, Kavala	7	36.5	47.7	1.7	1.5	na	na	na	5.4
MK4	Makrygialos	6.2	58.7	29.7	1.2	na	0.7	0.5	na	1.5
MK5	Makrygialos	13.2	43.4	32.0	2.3	na	2.2	na	na	4.5
MK6	Makrygialos	4.8	23.9	65.0	1.2	na	2.3	na	na	1.1
MK7	Makrygialos	4.6	9.4	86.0	na	na	na	na	na	na
MK10	Makrygialos	na	4.4	95.6	na	na	na	na	na	na
MK11	Makrygialos	6.2	56.0	31.7	0.7	na	1.5	na	na	1
MK12	Makrygialos	5.3	30.0	60.2	1.0	na	1.0	na	na	1.6

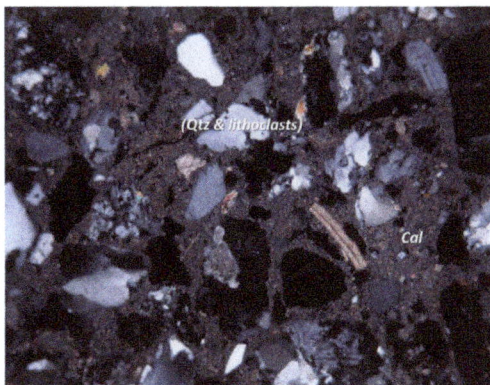

Figure 3. Crossed polar (XPL), magnification 5 × 5 mm, photograph length 2 mm. Byzantine mortar AN1b. The binder is composed microcrystalline calcite while the aggregates are composed of quartz grains and lithoclasts in sedimentary petrology.

3.1.2. Fortification Walls in Drama

SEM-EDX microstructural observations and thin section analysis revealed the morphology of DR2c (Figure 4) representative sample from fortification walls in Drama. Bulk analysis of oxides (Table 1) with SEM demonstrates that mortar samples from Drama are composed of 13.0% CaO and 57.6% of SiO_2. Point analysis of elements in each sample also revealed that Si occurred as the most abundant element in all SEM-EDX measurements. Minor elements detected were Al, Mg, Cl and Fe. Parallel to elemental analysis using SEM, thin section analysis in the polarized optical microscope demonstrates that the aggregates in the mortar samples from Drama are composed of quartz, calcite, plagioclase, potassium feldspar, biotite, muscovite with traces of granite, actinolite and tourmaline. The matrix is composed of fine grained calcite. When viewed in XPL sample DR2c has an overall brownish color, since this sample contains minerals and fragments of crushed bricks with maximum size 1 cm. Looking at photomicrograph (Figure 5), quartz demonstrates a coarse structure, at the same time appears a strong recrystallization matrix in saturation boundaries. XPL observations reveal the presence of muscovite together with quartz. The aggregate is formed from a pozzolanic additive of crushed and powdered ceramics plus a geological material such as volcanic rock with particles of variable composition. Muscovite is dispersed through calcitic binding material in the form of leafs with size up to 1mm. The presence of plagioclase is rare. Ceramic components are composed of muscovite, quartz and calcite. Elemental analysis also detected traces of chlorine in all samples, providing evidence for extensive deposition of salts. All samples demonstrate porous morphology that enhances the transportation of salts in the mortar structures using moisture. The black holes in the mortar structure are attributed to shrinkage within the polishing process and they do not represent evidence of porosity.

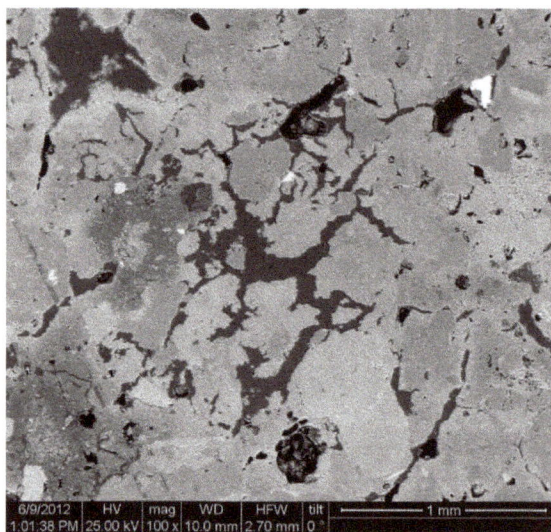

Figure 4. SEM/BSE (×253) photomicrograph of sample DR2c from fortification walls in Drama. Microstructural observations and thin section analysis.

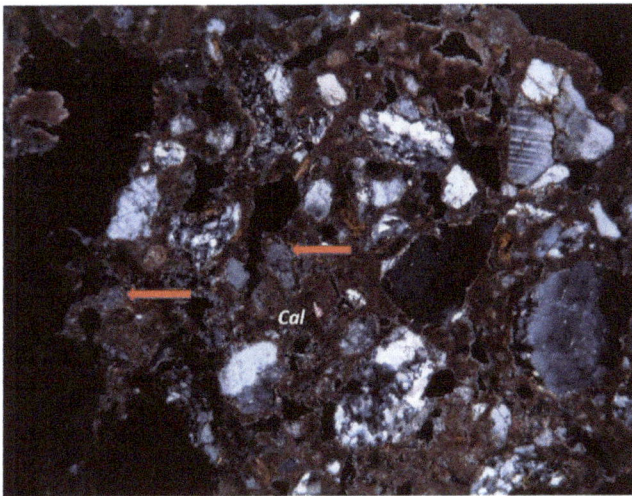

Figure 5. XPL, magnification 13 × 10 mm, +N, photograph length 1mm. Byzantine mortar, DR2c from fortification walls in Drama. The binding material is composed of calcite, pozzolanic reactions are observed (the arrows indicate the reactions), crushed ceramics and rock fragments compose the aggregates.

3.1.3. Marmarion Tower—Kavala

SEM/BSE and XPL photomicrographs demonstrate the microstructure of sample MA3 representative of samples obtained from Marmarion tower in Kavala (Figures 6 and 7). Bulk analysis of oxides shows that this sample contains 47.73% of CaO and 36.54% of SiO_2. Point analysis of individual elements showed that the aggregate is mostly composed of Si-rich particles while the matrix is composed of Ca-rich particles. The mineralogical composition of the matrix is detected using thin section analysis; the binding material is composed of fine-grained calcite. The aggregate can be readily discriminated from the matrix on the basis of its different mean atomic number, as shown in Figure 6. The aggregate is composed of quartz, calcite, plagioclase, potassium feldspar, microcline, biotite, muscovite while traces of chlorite were detected in the sample. XPL observations demonstrate that the length of quartz rises up to 2 mm and it is usually coarse. Quartz grains are surrounded by a strong recrystallization layer in the saturation boundaries. Calcite demonstrates intensive cracks and fragmentation, while signs of recrystallization layers are observed in a few samples. Elemental analysis also detected traces of Cl-rich particles in all samples, providing evidence for deposition of salts. All samples demonstrate porous morphology, which enhances the transportation of salts in the mortar structures using moisture.

Figure 6. SEM/BSE (×300) photomicrograph of sample MA3 obtained from Marmarion tower.

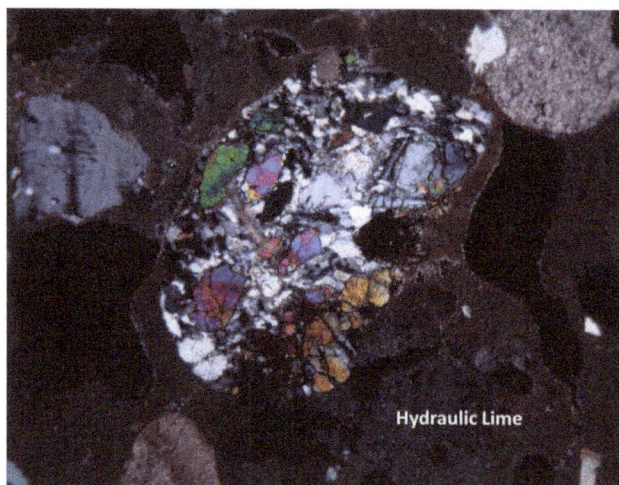

Figure 7. XPL, magnification 6 × 5 mm, +N, photograph length 2 mm. Sample MA3 obtained from Marmarion tower, it is a typical lime pozzolan mortar.

3.1.4. Makrygialos

Backscattered electron and XPL photomicrographs of sample MK4 (Figures 8 and 9), illustrate the morphology of typical mortar sample from Makrygialos, Greece. The photomicrographs (Figure 8a,b) depict large irregularly shaped lithoclast composed mainly of quartz (up to 58.7%) and feldspar, as shown by bulk analysis of oxides in the SEM (Figure 10). Petrographic observation elucidates that this is a textured mortar with mineral components sized from 100 μm to 4 mm. The matrix is composed from fine calcitic particles including also quartz grains, bulk analysis of oxides detected

95.6% percentage of CaO (Table 1, Figure 10). The matrix is composed of micritic calcite that is interrupted by angular aggregates and pores. XPL observation demonstrated that the mineralogical composition of the aggregate is: microcrystalline calcite with quartz, plagioclase, orthoclase, microcline, biotite, muscovite and traces of chlorite, zircon, titanite. Volcanic grains are also observed. The size of quartz grains is approximately 2 mm. The black holes in the pictures (Figure 8a,b) are pores in the mortar structure. In this sample, it is possible to have two types of pores. The elongate cracks are attributed to shrinkage and the rounded pores are derived from bubbles during setting. In places where heating was more intense, the half-burnt limestone fragments show an edge along the area of a crack (pore) [(Figure 8b, indicated with yellow arrow)]. It represents deposition of calcite on the edge of the crack due to the movement of solutions during burial. Thin section analysis also reveals the presence of unreacted lime and rare particles of partially or unburned limestone. The black hole in Figure 8b is lined with a very light grey line composed of calcite. This pore lining requires a source of dissolved carbonate. The dissolved carbonate could be attributed to the dissolution of calcite matrix or dissolution of the calcitic particles in the aggregate. In sample MK4 the matrix is mostly composed of calcite, therefore the pore lining is attributed to the dissolution of calcite matrix. Fine calcite particles (indicated with black arrow) in the matrix is attributed to limestone relicts that are derived from insufficient burning.

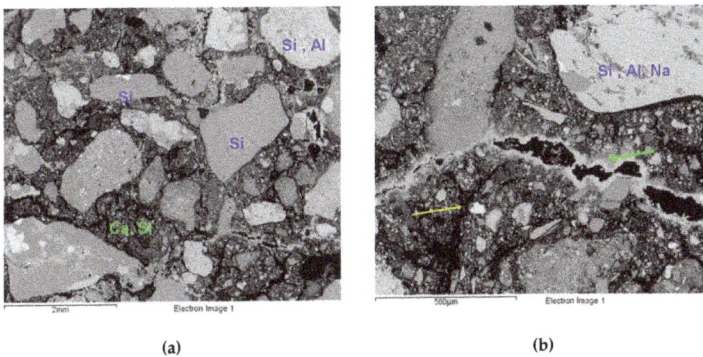

(a)   (b)

Figure 8. (**a,b**) SEM/BSE (×100) photomicrographs of sample MK4 obtained from Makrygialos. The potential minerals are quartz (up to 58.7%) and feldspar. The matrix is composed mainly from calcite and quartz. The arrows in (**b**) indicate half-burnt limestone fragments.

Figure 9. XPL, magnification 1 × 5 mm, +N, photograph length 2 mm. Hellenistic mortar sample MK4 obtained from Makrygialos, Greece. Microcrystalline calcite and plagioclase as aggregates.

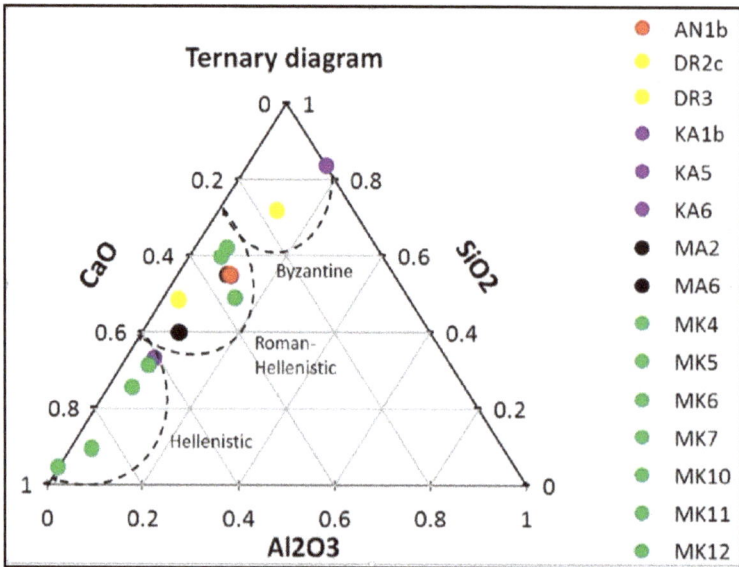

Figure 10. $CaO-Al_2O_3-SiO_2$% ternary plot of Energy Dispersive X-ray (EDX) analyses of data collected from mortar samples: Hellenistic mortars (MK4-12), Roman mortars (A1b, DR2c, DR3), Byzantine mortars (KA1β, K5, K3A).

3.2. Stable Isotope Analysis

The $\delta^{13}C$ and $\delta^{18}O$ of calcite matrix of the Byzantine, Late Roman and Hellenistic mortar indicate a wide range of isotopic values ranging between −17.1‰ to 1.2‰ and −25.9‰ to −2.0‰, respectively (Table 2). Additional samples of local marine limestone of Cretaceous age were obtained. These samples are located in the vicinity of the historical buildings and they were used as ancient raw materials for burning. The limestones comprise a range of $\delta^{13}C$ values between 0 and 3‰ and of $\delta^{18}O$ between −5‰ and −2‰ [19] indicating that the isotopic compositions of the calcite matrix from historical mortar are in general lighter, compared to those of local limestone. These light isotopic values are attributed to the absorption of carbon dioxide from the atmosphere within the setting of lime mortar ($^{13}C_{CO_2}$ = −6).

22 drill-core samples (presented in Table 2 and they represent samples with increasing isotopic values from the exterior to the interior layer) were examined in order to create an ideal Hellenistic and Byzantine mortar layer and to provide weathering gradients using isotopic analysis. The isotopic values of lime-pozzolan and lime mortar are shown in (Figure 11). In Table 3 we present the results of cross-section analysis. The isotopic values of these samples comprise a range of $\delta^{13}C$ and $\delta^{18}O$ values from −13.6 ‰ to 3.6‰ and from −12.6 ‰ to 0.4‰, respectively. Most of these samples indicate that calcite matrix is isotopically heavier inside the mortar layer, compared to those of the exterior layer, probably due to the reaction of atmospheric CO_2 with the lime mortar. A positive correlation of $\delta^{13}C$ vs. $\delta^{18}O$ of calcite is observed that is expressed by the regression lines $\delta^{18}O_{calcite\ matrix} = 0.61 \times \delta^{13}C_{calcite\ matrix} - 1.8$ (lime mortars line in Figure 11) and $\delta^{18}O_{calcite\ matrix} = 0.63 \times \delta^{13}C_{calcite\ matrix} - 2$ (hydraulic mortars line in Figure 11) for Hellenistic and Byzantine mortar, respectively. The slope of these equations is close to that is expressed by the experimentally obtained equation from Kosednar Legenstein et al., [20].

Table 2. List of analyzed samples and their isotopic value.

SampleID	Observation	$\delta^{13}C$ (‰ VPDB)	$\delta^{18}O$ (‰ VPDB)	Location
AN1a-1	Ext.Joint	−12.9	−16.1	Anaktoroupoli, Kavala
AN2a-1	Ext.Joint	−10.0	−9.2	Anaktoroupoli, Kavala
AN2a-2	Ext.Joint	−5.8	−5.5	Anaktoroupoli, Kavala
AN3a-1	Ext.Joint	−0.4	−2.4	Anaktoroupoli, Kavala
AN1b-1	Ext.Joint	−15.7	−12.5	Anaktoroupoli, Kavala
AN9-1	Ext.Joint	−8.7	−6	Anaktoroupoli, Kavala
Anp-1	Ext.Joint	−9.9	−4.9	Anaktoroupoli, Kavala
DRa-1	Ext.Joint	−10.0	−9.2	Drama
DRa-2	Ext.Joint	−17.1	−25.9	Drama
DR1b-1	Ext.Joint	−8	−13.4	Drama
DR2a-1	Ext.Joint	−8.9	−13.2	Drama
DR2c-1	Ext.Joint	−13.7	−16.1	Drama
DR5a-1	Ext.Joint	−12.9	−16.1	Drama
DR5b-1	Ext.Joint	−15.6	−10.5	Drama
DR2g-1	Ext.Joint	−6.6	−7.5	Drama
MA1-3	Ext.Joint	−10.9	−17.6	Marmarion Tower, Kavala
MA3-1	Ext.Joint	−8.3	−14.4	Marmarion Tower, Kavala
MA5-1	Ext.Joint	−5	−10.7	Marmarion Tower, Kavala
MA5-2	Int.joint	−8.3	−10	Marmarion Tower, Kavala
MK1-1	Rend.Ext.	−12.3	−6.3	Makrygialos
MK1-2	Rend.Int	−13.8	−9.1	Makrygialos
MK1-3	Rend.Int	−14.4	−9.5	Makrygialos
MK2-1	Rend.Ext.	−6.2	−4.6	Makrygialos
MK2-3	Rend.Int	−9.4	−7.3	Makrygialos
MK3-1	Rend.Ext.	−10.4	−8.7	Makrygialos
MK4-1	Rend.Ext.	−9.4	−6.4	Makrygialos
MK4-2	Rend.Int	−10.9	−9.1	Makrygialos
MK5-1	Rend.Ext.	−9.3	−5.7	Makrygialos
MK5-2	Rend.Int	−9.6	−7.1	Makrygialos
MK6-1	Rend.Ext.	0.1	−2.0	Makrygialos
MK6-2	Rend.Int	−9.9	−4.9	Makrygialos
MK7-1	Rend.Ext.	−5.0	−6.2	Makrygialos
MK7-2	Rend.Int	−8.7	−6.0	Makrygialos
MK8-1	Rend.Ext.	1.2	−2.1	Makrygialos
M8-2	Rend.Int	−3.0	−3.3	Makrygialos
MK9-1	Rend.Ext.	−0.4	−2.4	Makrygialos
MK9-2	Rend.Int	−10.6	−6.7	Makrygialos
MK10-1	Rend.Ext.	−5.8	−5.5	Makrygialos
MK10-2	Rend.Int	−3.0	−3.4	Makrygialos

Notes (Description of the type of mortar layer): Ext.joint = external joint; Int.joint = internal joint; Rend.Ext = rendering external; Rend.Int = rendering internal.

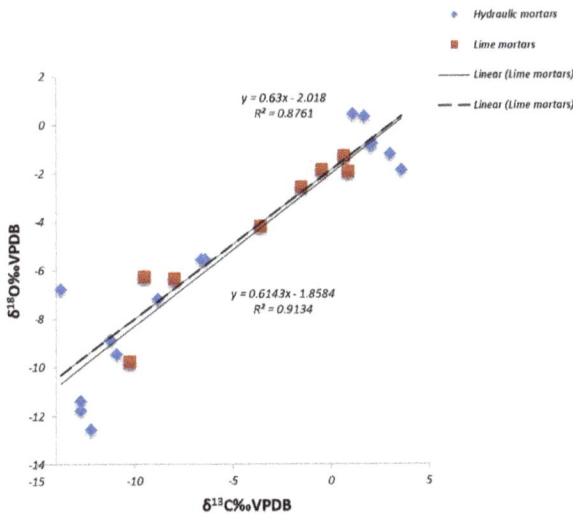

Figure 11. $\delta^{13}C$ vs. $\delta^{18}O$ for all the samples. Regression lines for hydraulic and lime mortars.

Table 3. List of cross-section analysis.

Sample ID	Depth (cm)	$\delta^{13}C$ (‰ VPDB)	$\delta^{18}O$ (‰ VPDB)
	−6	3.6	−1.9
	−6	3	−1.3
	−6	2	−0.9
	−6	1.7	0.3
	−6	1.1	0.4
	−3	−6.4	−5.6
Hydraulic. (lime-pozzolan) mortars.	−3	−6.6	−5.6
(Kavala. Drama)	−3	−8.8	−7.2
	+0	−10.9	−9.5
	+0	−12.2	−12.6
	+0	−11.2	−8.9
	+0	−12.7	−11.8
	+0	−12.7	−11.4
	+0	−13.6	−6.8
	−6	0.7	−1.3
	−6	0.9	−2
	−6	−0.5	−1.9
Lime mortars	−3	−1.6	−2.6
(Hellenistic mortars. Makrygialos)	−3	−3.6	−4.2
	+0	−8	−6.4
	+0	−9.5	−6.3
	+0	−10.3	−9.8

These lines comprise ideal mortar layers representing continues calcite formation with continues enrichment of ^{18}O versus ^{16}O and ^{13}C vs. ^{12}C in the CO_2 gas phase and the precipitated calcite. It appears that the quantity percentage and the quality of the charge do not influence significantly the ^{13}C of carbonate formed. The other data from mortars lie close or below this line indicating different setting environments and secondary effects. Also, in most Hellenistic samples it is noted that at interior levels the calcite is depleted in ^{13}C, compared to the middle layer. This depletion in ^{13}C can be explained by the heterogeneity of lime mortar, because this material is not chemically homogenous [36]. Also, higher $\delta^{18}O$ values are present in the near surface layer. The enrichment in ^{18}O in the near-surface layers is due to the contribution of water even if only 1/3 of the oxygen in the solid carbonate came from the water. Consequently, despite the contribution of water is minor, it causes this "surface effect" or oxygen enrichment because of important evaporation of water especially near the surface of the mortar (see Section 4.2 for further analysis).

4. Discussion

4.1. Technological Variations and Degradation Mechanisms

In comparing the technology of Hellenistic mortars from Makrygialos Greece with the later mortars (Roman and Byzantine) the first aspect to emphasize is that the lime-based mortars used in funerary monuments in Makrygialos were replaced with advanced lime-pozzolan mortars in Roman and Byzantine constructions. The content of lime in Hellenistic mortars approaches the 96%, whereas the content of lime in Roman and Byzantine mortars is between 30% and 90%. Furthermore, the aggregates in Hellenistic mortars are composed of quartz, plagioclase and potassium feldspar, whereas the aggregates of later mortars are composed of quartz, plagioclase, potassium feldspar, microcline, crushed ceramic and lithoclasts. The ternary plot (Figure 10) depicts the differentiations in mortar composition, since mortar samples obtained from Makrygialos are concentrated in the corner that expresses the maximum content of (CaO), moreover Roman and Byzantine mortars are concentrated in the corner with maximum content of (SiO_2). Roman and Byzantine mortars demonstrate coherence and strength since pozzolan is added to the binding material. The addition of pozzolanic materials enhances the power of the mortars and reduces the pore structure [37–39]. The addition of crushed ceramics as aggregates can trigger reactions on the fragment-lime interface [17,40] and the penetration of lime into the smaller pores of the ceramic fragments increases the apparent density which in turn increases the mortar strength, turning them suitable to support heavy historic constructions.

4.2. Evaluation of Setting Environments and Processes

The stable isotope (^{13}C and ^{18}O) analysis is related to the conditions of formation and the origin of the carbonates and therefore it is a useful tool for providing extra parameters, associated with the diagnostic of the mechanisms and processes that cause material degradation. The carbon and oxygen isotopes in mortar were analyzed in order to evidence the carbonate origin and to identify the potential sources of mortar decay such as salts attack, signs of sulfation and dissolution/recrystallization processes. All the mechanisms and processes related to mortar degradation are depicted in Figure 12. Figure 12 shows the various sources of CO_2 and H_2O (lines) and mechanisms (areas). line 1 is the equation $\delta^{18}O_{calcite} = 0.63 \times \delta^{13}C_{calcite} - 2$ and line 1a is the deviation from line 1 due to relics of local limestone used for burning or contamination by limestone aggregates; line 2a the primary source of water used for setting of the lime mortar is originating from heavy source—evaporation effect. Point CM indicates that precipitated calcite is formed directly by the absorption of atmospheric CO_2 in strong alkaline aqueous environment; its values change to more enriched isotopic composition due to continuous enrichment of $\delta^{13}C$ of CO_2 during calcite precipitation; area A: defined by 1 and 1a lines, precipitated calcite formed from atmospheric CO_2 and contaminated by residual natural limestone; area B indicates that precipitated calcite is formed from atmospheric CO_2 and heavy, evaporated water. Enrichment of C and O isotopes indicates recrystallization of calcite with water of meteoric origin and atmospheric CO_2; area B2 these samples demonstrate enrichment in ^{18}O of calcite due to equilibrium to silicate minerals.; area C precipitated calcite formed from atmospheric CO_2 and isotopically light local meteoric water or isotopically light recondensed primary waters.; area C1, depletion of C and O isotopes, indicates also human influence (surface treatment) and biological growth; Figure 13 presents a simplification of Figure 12. The blue arrow indicates the enrichment of ^{18}O of calcite due to the equilibrium with the silica mineral, whereas the red arrow denotes the enrichment of ^{18}O of precipitated calcite formed from atmospheric CO_2 and heavy evaporated water.

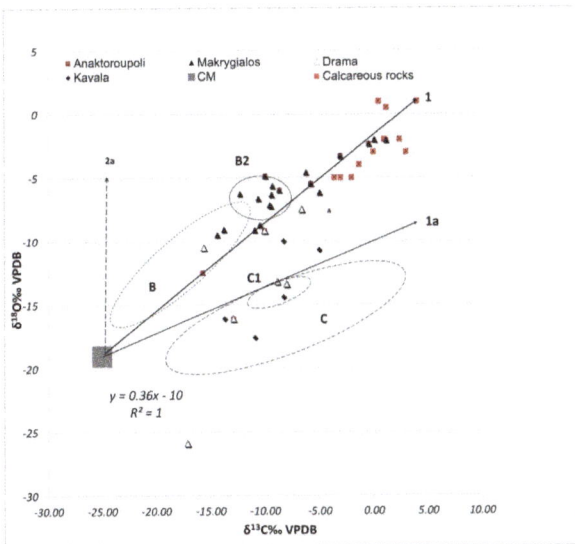

Figure 12. Scattered diagram that summarizes the various sources of CO_2 and H_2O (lines) and mechanisms (areas).

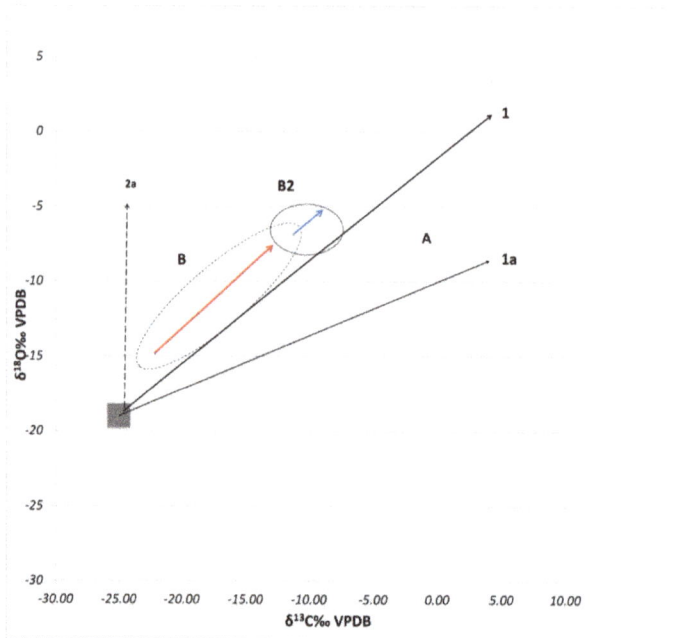

Figure 13. The blue arrow indicates the enrichment of ^{18}O of calcite due to the equilibrium with the silica mineral, whereas the red arrow denotes the enrichment of ^{18}O of precipitated calcite formed from atmospheric CO_2 and heavy evaporated water.

The isotopic composition of calcite from mortar represent non-isotopic equilibrium [19,20,41–45] and depends mainly on the isotopic composition of atmospheric CO_2 and water and on the degree to which the isotopic equilibrium is reached and therefore the factor of fractionation of $\delta^{13}C_{CaCO_3}$ $-\delta^{13}C_{CO_2} -\delta^{18}O_{CO_2}$ and $\delta^{18}O_{H_2O}$. In order to define the isotopic values for a pure calcite precipitated in alkaline environment using local meteoric water and atmospheric CO_2, the isotopic values of local water in each sampling area are analyzed. The isotopic values of rain and spring water of the area where the ancient monuments are located are: $-7.2‰$ ^{18}O and $-45‰$ ^{2}H for samples from Makrygialos, $-6.5‰$ ^{18}O and $-38‰$ ^{2}H for samples from Kavala and $-8‰$ ^{18}O and $-65‰$ ^{2}H for samples from Drama [46]. Moreover, evaporation of water during the setting of the mortar gives isotopically heavier residual water. In contrary condensation effect gives isotopically lighter residual water. The origin of atmospheric CO_2 for calcite formation is another important factor. It can be presumed that the isotopes of C and O of CO_2 (of atmospheric origin) were stable for the last two decades. However, air pollution in not a new problem. In ancient Rome glassworkers moved outside the city centre because of environmental nuisance [47]. CO_2 of biogenic origin, (combustion of coal, fuel or soil CO_2) presents very low $\delta^{13}C$ [19].

Therefore, considering the isotope fractionation of C between the CO_2 (gas) and the precipitated calcite in alkaline environment the $\delta^{13}C_{calcite-CO_2} = \delta^{13}C_{calcite} - \delta^{13}C_{CO_2} = -18‰$ [20] and considering that the values for $\delta^{13}C_{CO_2}$ range between $-7‰$ and $-9‰$, it is possible to calculate the $\delta^{13}C_{calcite}$ that is between $-25‰$ and $-27‰$. Regarding the $\delta^{18}O$, both the water and the CO_2 oxygen contribute to the precipitated carbonate. Usdowski and Hoefs [48] and O'Neil and Barnes [41] have demonstrated that in alkaline environments, 2/3 of the oxygen in precipitated carbonate comes directly from CO_2 and 1/3 of the $\delta^{18}O$ comes from OH^-. Considering the mean temperature in Makrygialos and East Macedonia and considering also the isotope fractionation between H_2O and OH^- ($\alpha_{H_2O - OH^-} = 1.042$ $- 1$, T = 20 °C), the value of $\delta^{18}O_{OH^-}$ [$\delta^{18}O_{OH^-} = \delta^{18}O_{H_2O} - (\alpha_{H_2O - OH^-} - 1) \times 1000$] is calculated between −78‰ and −80‰. Taking into consideration that the $\delta^{18}O$ of CO_2 is 10‰, (VPDB), the $\delta^{18}O_{calcite}$ is calculated ($\delta^{18}O_{calcite} = 1/3 \times \delta^{18}O_{OH^-} + 2/3 \times \delta^{18}O_{CO_2}$) between −19‰ and −20‰. Therefore, such C and O isotopes values $\delta^{13}C_{calcite} = -25‰$ to −27‰ and $\delta^{18}O_{calcite} = -19‰$ to −20‰ (area CM in Figure 12) are typical for the precipitation of calcite by atmospheric CO_2 absorption using the local isotopic composition of Drama, Kavala and Makrygialos water and atmospheric CO_2.

However, the carbon and oxygen isotopic value could change due to continuous enrichment of ^{13}C relative to ^{12}C and of ^{18}O relative to ^{16}O for the CO_2 gas phase and the precipitated calcite due to kinetic isotope fractionation, during calcite formation. This positive correlation of $\delta^{13}C/\delta^{18}O$ of the formed calcite in our historical mortar made possible to determine the ideal layer from Hellenistic period and is represented by the equation $\delta^{18}O_{calcite} = 0.63 \times \delta^{13}C_{calcite} - 2$ (line 1 in Figure 12). Moreover, the $\delta^{13}C$ of mortar is also affected by the mortar mixing composition (the parts of mixing between lime, water, sand and old calcite or aggregate). If the lime is pure CaO (100%), then the carbonate formed, still has not attained a steady isotopic composition and presents $\delta^{13}C$ and $\delta^{18}O$ −22‰ and −20‰, respectively. These isotopic values change as the amount of charge increases (because the $\delta^{13}C$ of old calcite is approximately 0‰) and the $\delta^{13}C$ of carbonate formed depends on the mixing percentage of CaO and sand-old calcite. The percentage of CaO in Hellenistic samples is between 50% and 95% (Table 1). So, for a mix of equal quantities of CaO (50%) and sand-old calcite (50%), the $\delta^{13}C$ is −9.5‰ while, when the percentage of CaO is higher due to the addition of small quantities of sand and old calcite, the $\delta^{13}C$ of carbonate formed is more negative. Additionally, local marine limestone (Cretaceous carbonate) in the vicinity of the historical buildings, which was the main source of ancient raw material for burning, has $\delta^{13}C$ between 0 and 3‰ and $\delta^{18}O$ between −5‰ and −2‰ (Figure 12). Therefore, it is expected that the old mortars, which are a mixture of lime, water, sand and an aggregate like reworked brick and marble, would have $\delta^{13}C > -25‰$ or heavier depending on the percentage of participation of old calcite. Taking into account the percentage of CaO in Hellenistic samples the $\delta^{13}C$ of carbonate formed will be between −25‰ and −9.5‰ depending on the participation of old calcite aggregates. In fact the Hellenistic samples show a positive correlation between $\delta^{13}C$ and charge indicating that the shift, to more positive values can be caused by residual limestone. Consequently, the percentage of limestone used for burning or the CO_2 absorption is competitive against the final value of ^{13}C (lines 1 and 1a, Figure 12).

Mineralogical, morphological and chemical analysis indicated that the hydraulic mortars are attacked by salts while the lime-based mortars are decomposed because of the leaching action that causes material dissolution/recrystallization processes. This is depicted in Figure 12, since samples from Makrygialos are concentrated in area B2; $\delta^{18}O$ values of these samples demonstrate continues enrichment and recrystallization tendency in O, which is related to the changes in isotopic composition because of recrystallization/dissolution mechanisms. Additionally, most of the data lie close or between on these lines indicating that the isotopic values reflect increasing $\delta^{13}C$ values of CO_2 with calcite formation along the mortar layer and the contamination by impurities of limestone relict.

However, a certain number of data (samples from Marmarion MA1-3, MA5-1 and Drama DR2g-1, DR2a-1, samples from Makrygialos and calcareous rocks) lie on or beyond of this area (area A defined by 1 and 1a lines) indicating different setting environment and secondary effects. A moderate oxygen isotopic shift is observed for samples from Makrygialos indicating that the variability of ancient water influenced the isotopic composition of calcite. Also, the oxygen values of the precipitated calcite shift to heavier values because of the evaporation of water (line 2a, area B) during the setting of mortar. Moreover, the mineralogical, morphological and chemical analysis showed that samples from Anaktoroupoli (Anp-1), Drama (DR5b-1) and some from Makrygialos (MK10-2, MK5) demonstrate intensive fragmentation, cracks and extensive deposition of salts that indicates later calcite alteration with actual meteoric water. There seems to be a moderate enrichment of $\delta^{18}O$ at the surface indicating influence from rain evaporated water and from capillary suction of water. In this case the capillary transport of water was followed by evaporation when equilibrium between capillary addition of water and evaporation is reached and by salt precipitation. Thus, the major processes are dissolution of calcite and probably precipitation of other minerals (Area B2). Most of the samples lying in area B2 come from Makrygialos. SEM analysis for these samples elucidated evidence of recrystallization and material dissolution. The rest of the samples (mainly from Makrygialos) present positive oxygen isotopic shift and also significant change of the initial carbon. These isotopic values may be caused by the recrystallization of calcite with porewater and CO_2 of various sources. Secondary solutions for recrystallization will be rain water and then the $\delta^{18}O$ of calcite matrix should shift to heavier values, in our case around 0‰ similar to isotopic values of limestone. Values around 0‰, also for oxygen and carbon are attributed in the fine calcite particles (according to mineralogical, morphological and chemical analysis) in the matrix due to limestone relicts that are derived from insufficient burning. A group of data lie beyond area A indicating that precipitated calcite is formed from atmospheric CO_2 and isotopically light local meteoric water or isotopically light re-condensed primary water (area C); also line 2a depicts the depletion of both isotopes indicating recrystallization of calcite with light water and CO_2 of mixed origin (atmospheric and soil origin).

5. Conclusions

This work indicated that the use of isotope methods in addition to the existing "toolbox" (mineralogical, morphological, chemical analysis) of pre-restoration research methods will help to better direct safeguarding and conservation measures. The diagnostic contribution of stable isotope study traces the various sources and possible secondary processes that are responsible for the mortar degradation. Stable isotope analysis applied in combination with mineralogical, microscopic and elemental analysis provided information on the different types of mortar degradation. Mortar's weathering was attributed to a combination of environmental threats such as salts attack and leaching action. Stable isotope analysis (^{13}C and ^{18}O) provided information relative to the origin of CO_2 and water during calcite formation making possible to distinguish different mortar technologies and degradation gradients. Compositional and morphological analyses were achieved using energy dispersive X-ray analysis in the scanning electron microscope while the mineralogical phases were detected using petrographic (polarized optical microscopy) analysis. The results of micro-morphological and petrographic examination elucidated the technological continuity and degradation of historic mortars. Hellenistic mortars are composed of lime enhanced with quartz aggregates. Roman and Byzantine mortars are composed of hydraulic lime, pozzolan and a various aggregates such as quartz, feldspar, ceramic and lithoclasts. The main degradation mechanisms are calcite recrystallization, loose of adhesion bonds in the binding material and salts crystallization. Mortar's weathering was also attributed to a combination of environmental threats such as salts attack and leaching action. Stable isotopes of $\delta^{18}O$ and $\delta^{13}C$ showed that in general there are extended dissolution/reprecipitation processes that take place on the surface layers ending in the inner structure of funerary monuments. Finally, this study indicated that stable isotope analysis is an excellent tool to fingerprint the origin of carbonate and therefore indicate the variations in mortar's technology, the environmental setting conditions of mortar, origin of CO_2 and water during calcite formation and to determine the weathering depth and the potential secondary degradation mechanisms.

Author Contributions: Conceptualization, D.E; Methodology, D.E, K.D., C.V. and D.G. Writing–Original Draft Preparation, D.E., K.D.; Writing-Editing, D.G.; Project Administration, D.E.; Funding Acquisition, D.E.

Funding: This research received no external funding.

Acknowledgments: We would like to sincerely thank Bessios at the KZ' Ephoreia of Classical and Prehistoric antiquities for kindly providing access for sampling at Hellenistic funerary monuments. Also, Dadaki and Ioannis Iliadis at the 12th Ephoreia of Byzantine antiquities are kindly thanked for providing Byzantine and Roman mortar samples. Finally, many thanks go to Vasilios Melfos for performing petrographic analysis in mortar samples at the Department of Mineralogy, Petrology and Economic Geology in Aristotle University of Thessaloniki.

Conflicts of Interest: The authors declare no conflict of interest.

Geosciences **2018**, *8*, 339

References

1. Sabbioni, C.; Zappia, G.; Riontino, C.; Blanco-Varela, M.T.; Aguilera, J.; Puertas, F.; Van Balen, K.; Toumbakari, E.E. Atmospheric deterioration of ancient and modern hydraulic mortars. *Atmos. Environ.* **2001**, *35*, 539–548. [CrossRef]

2. van Balen, K.; Toumbakari, E.E.; Blanco-Varela, M.T.; Aguilera, J.; Puertas, F.; Palomo, A. Environmental deterioration of ancient and modern hydraulic mortars. *WIT Trans. Built Environ.* **1970**, *42*, 10.

3. Yates, T. *The Effects of Air Pollution on the Built Environment. Air Pollution Reviews*; Brimblecombe, P., Ed.; Imperial College Press: London, UK, 2003; pp. 107–132.

4. Zappia, G.; Sabbioni, C.; Pauri, M.; Gobbi, G. Mortar damage due to airborne sulfur compounds. *Mater. Struct.* **1994**, *27*, 469–473. [CrossRef]

5. Stefanidou, M.; Papayianni, I. Salt accumulation in historic and repair mortars. In Proceedings of the Heritage, Weathering and Conservation Conference, Consejo Superior de Investigaciones Cientificas, Madrid, Spain, 21–24 June 2006; Taylor & Francis: London, UK, 2006; pp. 269–272.

6. Stefanidou, M. A contribution to salt crystallization into the structure of traditional repair mortars through capillarity. In Proceedings of the 11th euroseminar on microscopy applied to building materials, Porto, Portugal, 5–9 June 2007.

7. Theoulakis, P.; Moropoulou, A. Microstructural and mechanical parameters determining the susceptibility of porous building stones to salt decay. *Constr. Build. Mater.* **1997**, *11*, 65–71. [CrossRef]

8. Rossi-Manaresi, R.; Tucci, A. Pore structure and the disruptive or cementing effect of salt crystallization in various types of stone. *Stud. Conserv.* **1991**, *36*, 56–58.

9. Scherer, G.W. Stress from crystallization of salt. *Cem. Concr. Res.* **2004**, *34*, 1613–1624. [CrossRef]

10. Papayianni, I.; Stefanidou, M.; Pachta, V.; Konopisi, S. Content and topography of salts in historic mortars. In Proceedings of the 3rd Historic mortars conference, Glasgow, Scotland, 11–14 September 2013.

11. Gomides, M.d.J.; Molin, D.C.C.D.; Rêgo, J.H.d.S. Effect of the incorporation of aggregates with high sulfide content on the mechanical and microstructural properties of concrete with slag cement. *Matéria (Rio de Janeiro)* **2017**, *22*. [CrossRef]

12. Chinchón, J.; Ayora, C.; Aguado, A.; Guirado, F. Influence of weathering of iron sulfides contained in aggregates on concrete durability. *Cem. Concr. Res* **1995**, *25*, 1264–1272. [CrossRef]

13. Clark, I.; Fritz, P. *Environmental Isotopes in Hydrogeology*; Lewis Publishers: Boca Raton, FL, USA, 1997.

14. Craig, H. Isotopic variations in meteoric waters. *Science* **1961**, *133*, 1702–1703. [CrossRef] [PubMed]

15. Maravelaki-Kalaitzaki, P.; Bakolas, A.; Moropoulou, A. Physico-chemical study of Cretan ancient mortars. *Cement and Concrete Research* **2003**, *33*, 651–661. [CrossRef]

16. Moropoulou, A.; Bakolas, A.; Bisbikou, K. Investigation of the technology of historic mortars. *J. Cult. Her.* **2000**, *1*, 45–58. [CrossRef]

17. Moropoulou, A.; Bakolas, A.; Bisbikou, K. Characterization of ancient, byzantine and later historic mortars by thermal and X-ray diffraction techniques. *Thermochim. Act.* **1995**, *269*, 779–795. [CrossRef]

18. Bakolas, A.; Biscontin, G.; Moropoulou, A.; Zendri, E. Characterization of the lumps in the mortars of historic masonry. *Thermochim. Acta* **1995**, *269*, 809–816. [CrossRef]

19. Dotsika, E.; Psomiadis, D.; Raco, B.; Poutoukis, D.; Gamaletsos, P. Isotopic analysis for degradation diagnosis of calcite matrix in mortar and plaster. *Anal. Bioanal. Chem.* **2009**, *395*, 2227–2234. [CrossRef] [PubMed]

20. Kosednar-Legenstein, B.; Dietzel, M.; Leis, A.; Stingl, K. Stable carbon and oxygen isotope investigation in historical lime mortar and plaster–Results from field and experimental study. *Appl. Geochem.* **2008**, *23*, 2425–2437. [CrossRef]

21. Larbi, J. Microscopy applied to the diagnosis of the deterioration of brick masonry. *Constr. Build. Mater.* **2004**, *18*, 299–307. [CrossRef]

22. Hughes, J.J.; Cuthbert, S.J. The petrography and microstructure of medieval lime mortars from the west of Scotland: Implications for the formulation of repair and replacement mortars. *Mater. Struct.* **2000**, *33*, 594–600. [CrossRef]

23. Deer, W.A.; Howie, R.A.; Zussman, J. *An Introduction to the Rock Forming Minerals*, 2nd ed.; Longman Scientific & Technical: London, UK, 1992; p. 696.

24. Tucker, M.E. *Sedimentary Petrology*, 2nd ed.; Blackwell Scientific Publications: Hoboken, NW, USA, 1991.

25. Adams, A.E.; MacKenzie, W.S.; Guildford, C. *Atlas of Sedimentary Rocks Under the Microscope*; Routledge: Abington, UK, 2017.

26. Whitbread, I.K. The characterisation of argillaceous inclusions in ceramic thin sections. *Archaeometry* **1986**, *28*, 79–88. [CrossRef]

27. Freestone, I.C. Ceramic Petrography. *Am. J. Archaeol.* **1995**, *99*, 111–115.

28. Coplen, T.B; Kendall, C.; Hopple, J. Comparison of stable isotope reference samples. *Nature* **1983**, *302*, 236–238. [CrossRef]

29. Coplen, T.B. Reporting of stable hydrogen, carbon and oxygen isotopic abundances (technical report). *Int. Union Pure Appl. Chem.* **1994**, *66*, 273–276. [CrossRef]

30. Coplen, T.B. Discontinuance of Smow and PDB. *Nature* **1995**, *375*, 285. [CrossRef]

31. Chabas, A.; Jeannette, D. Weathering of marbles and granites in marine environment: petrophysical properties and special role of atmospheric salts. *Environ. Geol.* **2001**, *40*, 359–368. [CrossRef]

32. Liu, P. Damage to concrete structures in a marine environment. *Mater. Struct.* **1991**, *24*, 302–307. [CrossRef]

33. Tsouris, E.M.A. Byzantine fortifications in Evros. *Byzantina* **2006**, *26*, 153–209.

34. Dadaki, S. (Greek Ministry of Culture, Athens, Greece). Historical overview of the antiquities of 12th EBA, 12th Ephoreia of Byzantine antiquities, Archaeological report. Unpublished work, 2010.

35. Mpesios, M. *Pieridon Stephanos: Pydna, Methone Kai Hoi Archaiotetes Tes Voreias Pierias*; Anthropon Physeos Erga: Katerine, Greece, 2010.

36. van Strydonck, M.J.; Dupas, M.; Keppens, E. Isotopic fractionation of oxygen and carbon in lime mortar under natural environmental-conditions. *Radiocarbon* **1989**, *31*, 610–618. [CrossRef]

37. Papayianni, I.; Stefanidou, M. Strength–porosity relationships in lime–pozzolan mortars. *Constr. Build. Mater.* **2006**, *20*, 700–705. [CrossRef]

38. Walker, R.; Pavía, S. Behaviour and Properties of Lime-Pozzolan Pastes. In Proceedings of the 8th International Masonry Conference 2010, Dresden, Germany, 4–7 July 2010; Jäger, W., Haseltine, B., Fried, A., Eds.; International Masonry Society: Shermanbury, UK, 2010; pp. 353–362.

39. Walker, R.; Pavía, S. Physical properties and reactivity of pozzolans, and their influence on the properties of lime-pozzolan pastes. *Mater. Struct.* **2011**, *44*, 1139–1150. [CrossRef]

40. Matias, G.; Faria, P.; Torres, I. Lime mortars with heat treated clays and ceramic waste: a review. *Constr. Build. Mater.* **2014**, *73*, 125–136. [CrossRef]

41. O'Neil, J.R.; Barnes, I. ^{13}C and ^{18}O compositions in some fresh-water carbonates associated with ultramafic rocks and serpentinites: Western United States. *Geochim. Cosmichim. Acta.* **1971**, *35*, 687–697. [CrossRef]

42. Pachiaudi, C.; Marechal, J.; van Strydonck, M.; Dupas, M.; Dauchot- Dehon, M. Isotopic fractionation of carbon during CO_2 absorption by mortar. *Radiocarbon* **1986**, *28*, 691–697. [CrossRef]

43. Rafai, N.; Letolle, R.; Blanc, P.; Gegout, P.; Revertegat, E. Carbonation–decarbonation of concretes studied by the way of carbon and oxygen stable isotopes. *Cem. Concr. Res.* **1992**, *22*, 882–890. [CrossRef]

44. Dietzel, M.; Usdowski, E.; Hoefs, J. Chemical and $^{13}C/^{12}C$and $^{18}O/^{16}O$-isotope evolution of alkaline drainage waters and the precipitation of calcite. *Appl. Geochem.* **1992**, *7*, 177–184. [CrossRef]

45. Clark, I.D.; Fontes, J.C.; Fritz, P. Stable isotope disequilibria in travertine from high pH waters: Laboratory investigations and field observations from Oman. *Geochem. Cosmchim. Acta* **1992**, *56*, 2041–2050. [CrossRef]

46. Dotsika, E.; Lykoudis, S.; Poutoukis, D. Spatial distribution of the isotopic composition of precipitation and spring water in Greece. *Glob. Planet. Chang.* **2010**, *71*, 141–149. [CrossRef]

47. Taylor, R.M. *Roman Builders: A Study in Architectural Process*; Cambridge University Press: Cambridge, UK, 2003; pp. 175–178.

48. Usdowski, E.; Hoefs, J. Oxygen isotope exchange between carbonic acid, bicarbonate, carbonate, and water: A re-examination of the data of McCrea (1950) and an expression for the overall partitioning of oxygen isotopes between the carbonate species and water. *Geochim. Cosmochim. Acta* **1993**, *57*, 3815–3818. [CrossRef]

geosciences

MDPI

Article

Adapting Cultural Heritage to Climate Change Risks: Perspectives of Cultural Heritage Experts in Europe

Elena Sesana [1], Alexandre S. Gagnon [1], Chiara Bertolin [2] and John Hughes [1,*]

[1] School of Computing, Engineering and Physical Sciences, University of the West of Scotland, High St, Paisley PA1 2BE, UK; elena.sesana@uws.ac.uk (E.S.), alexandre.gagnon@uws.ac.uk (A.S.G.)
[2] Department of Architectural Design History and Technology, Norwegian University of Science and Technology (NTNU), Alfred Getz vei 3, 7491 Trondheim, Norway; chiara.bertolin@ntnu.no
* Correspondence: John.Hughes@uws.ac.uk; Tel.: +44-(0)-141-848-3268

Received: 6 July 2018; Accepted: 8 August 2018; Published: 14 August 2018

Abstract: Changes in rainfall patterns, humidity, and temperature, as well as greater exposure to severe weather events, has led to the need for adapting cultural heritage to climate change. However, there is limited research accomplished to date on the process of adaptation of cultural heritage to climate change. This paper examines the perceptions of experts involved in the management and preservation of cultural heritage on adaptation to climate change risks. For this purpose, semi-structured interviews were conducted with experts from the UK, Italy, and Norway as well as a participatory workshop with stakeholders. The results indicate that the majority of interviewees believe that adaptation of cultural heritage to climate change is possible. Opportunities for, barriers to, and requirements for adapting cultural heritage to climate change, as perceived by the interviewees, provided a better understanding of what needs to be provided and prioritized for adaptation to take place and in its strategic planning. Knowledge of management methodologies incorporating climate change impacts by the interviewees together with best practice examples in adapting cultural heritage to climate change are also reported. Finally, the interviewees identified the determinant factors for the implementation of climate change adaptation. This paper highlights the need for more research on this topic and the identification and dissemination of practical solutions and tools for the incorporation of climate change adaptation in the preservation and management of cultural heritage.

Keywords: adaptation; climate change; cultural heritage; management; conservation

1. Introduction

Our tangible cultural heritage is threatened by gradually shifting weather patterns and extreme events. An increase in temperature together with changes in precipitation, relative humidity, and wind, for instance, can negatively impact on the materials comprising cultural heritage assets. This is because a change in average climatic conditions as well as changes in the frequency and intensity of severe weather events can affect the biological, chemical, and physical mechanisms leading to degradation of the assets [1–5]. This includes an increase in the freeze-thaw cycle in northern Europe, extreme heat and droughts in the Mediterranean region, the overall decrease in summer precipitation in Europe, and an increase in winter storms and heavy precipitation events in the Atlantic region. In addition, cultural heritage sites in coastal regions are particularly at risk of sea-level rising (SLR) and the occurrence of storm surges while natural hazards such as floods, landslides, earthquakes, volcanoes, and fire can also have devastating impacts on cultural heritage assets. As the outputs from global climate models project that climatic changes will grow larger over the current century with the magnitude of the projected change dependent on the selected path of greenhouse gas (GHG) emissions and the model selected, strategies need to be developed to reduce the negative consequences of climate change on sites of historical value in addition to mitigate climate change by curtailing GHG emissions.

The United Nations Framework Convention on Climate Change (UNFCCC) established the need to address both mitigation and adaptation to climate change [6]. Hence, there are two strands to the climate change challenge. Mitigation encompasses measures and activities aimed at reducing GHG emissions or enhancing the sinks of such gases, whilst adaptation refers to any adjustments in a system in response to the actual or projected climatic stimuli [7,8], including changes in socio-environmental processes, perceptions, practices, and actions to reduce potential damages or to take advantage of new opportunities that may arise [9]. Mitigation has traditionally been given more attention in climate change research and policy than adaptation [10–13]. Nonetheless, there has been increasing interest in adaptation research since the late 1990s due to the recognition that the climate is already changing and that adaptation to the already unavoidable impacts is crucial [13,14]. In the field of cultural heritage, mitigation can involve improving the energy efficiency of historical buildings, for instance, while building a sea wall to protect coastal heritage sites from storm surges and SLR is an example of adaptation to climate change risks.

This paper focuses on adapting cultural heritage to climate change risks and is limited to the immovable and tangible cultural heritage, for example, historical buildings, monuments, and archaeological sites. The objectives of this paper are as follows:

- To understand how climate change adaptation is considered by experts in cultural heritage preservation in Europe. This includes determining the perspectives of these experts specifically on the requirements for, and opportunities and barriers to, adaptation, and the determinant factors for the implementation of adaptation efforts.
- To identify current examples of best practice in the management and practical site-level strategies and methodologies for adapting cultural heritage to climate change.

2. Adapting Cultural Heritage to Climate Change Risks

Previous studies have focused on assessing the impacts of climate change on cultural heritage sites in Europe [1–5,15], with limited research on adaptation. Efforts in the area of adaptation are yet limited to the dissemination of guidelines and recommendations for implementing adaptation measures [1,16–25], the identification of the determinants of adaptive capacity [26], and the identification of the barriers to adaptation [27–30]. Sabbioni et al. [1,16] developed guidelines for adapting the European cultural heritage to climate change impacts, which were later adopted by the Italian Strategic Agenda [31,32]. These included strategies for both physical adaptation and for adjusting management practices. Examples of the latter include improving the monitoring, maintenance, and preparedness to floods and landslides at cultural heritage sites [1,16]. Heathcote et al. [17] described the adaptive measures suggested in the Historic England climate change adaptation plan to cope with the impacts of climate change on cultural heritage in England, for example, developing approaches to deal with change and loss. Haugen and Mattson [18] investigated the impacts of climate change on cultural heritage in Norway, and recommended adaptive measures to deal with the risks identified. Also in Norway, Grøntoft [23] discussed options for adapting the surface of heritage material and the facades of heritage buildings to the impacts of climate change, including encouraging the adoption of adaptive measures that preserve building surfaces, incorporating climate change projections in building regulations, and developing new technologies to adapt buildings to future climatic conditions.

In addition to providing adaptive solutions to climate change impacts, some studies have examined the determinants of adaptive capacity in the field of cultural heritage, the role of tourism in supporting adaptation efforts, and key issues associated with adaptation of cultural heritage to climate change [25,26]. Phillips [26] investigated the adaptive capacity to climate change at cultural heritage sites and identified access to information, authority, resources, cognitive factors, leadership, and learning capacity as the key determinants. Hall [25] and Hall et al. [24] gave insight into the role of tourism in adapting cultural heritage to climate change. Furthermore, Hall et al. [24] identified key themes in dealing with the consequences of climate change for cultural heritage, such as the importance of using integrated approaches in adaptation and preserving the values interconnected with the local heritage through consultation with stakeholders that live and work within heritage sites, the need for a unified approach in cultural heritage preservation, and dealing with the paucity of funding for climate change adaptation.

Other studies have analysed climate change adaptation strategies for specific heritage typologies such as archaeological sites and historical districts located in coastal regions. A recurring theme emerging from these studies is the need to strengthen monitoring and maintenance and increase risk-preparedness and the dissemination of know-how when considering adaptation of cultural heritage to climate change. Cassar [19] investigated the impacts of climate change on archaeological sites and suggested adopting solutions that are sensibly designed to the specific conditions of the site after a long-term programme of monitoring and maintenance. She also deliberated on the adaptive solution adopted for the Megalithic Temples, a World Heritage Site (WHS) located in Malta. Additionally, Cassar [19] summarised the adaptation measures suggested by the UN Educational, Scientific and Cultural Organization (UNESCO) and by the International Council On Monuments and Sites (ICOMOS), who recommend increasing research, knowledge, education, engagement, the upgrading of management plans—including risk assessments—and monitoring procedures to increase the resilience of the sites. Pollard-Belsheim et al. [20] investigated the effectiveness of existing adaptation strategies to preserve coastal archaeological sites with a focus on a range of adaptation solutions such as watertight barriers, wooden breakwaters, and gabion rock wall. Climate change adaptation solutions to archaeological sites were also studied by Carmichael et al. [21] in indigenous communities in Australia. In the Republic of Tartasan in Russia, Usmanov et al. [33] suggested adaptation measures to preserve coastal archaeological sites from coastal erosion such as building breakwaters, changes in land use and planting trees. Nicu [34] also suggested planting trees to stabilise slopes as an adaptation measure to preserve the archaeological and paleontological sites of northeastern Romania, a region susceptible to landslides, a natural hazard whose occurrence could be affected by climate change. Fatorić and Seekamp [22] examined an approach to support decision-making by promoting climate change adaptation through the sharing of information and better engagement amongst stakeholders in the historical districts of the coast of North Carolina in the United States (US).

Until now, less research has been conducted on the identification of the barriers to adapting our cultural heritage to climate change [27–30] and in considering geological hazards in the assessment of vulnerability to inform the adaptation process. Phillips [27] investigated whether climate change is considered in the management plans of WHS in the United Kingdom (UK) and identified issues arising from its consideration, notably the lack of detailed information from climate change scenarios and the uncertainty associated with them, as well as the lack of resources, knowledge, and skills available. Also in the UK, it was noted that knowledge of geological and geomorphological processes can improve our understanding of the risk of natural hazards on WHS and thereby inform adaptation to climate change [35]. However, natural processes have often been ignored in previous assessments of vulnerability and the inclusion of geological hazards such as landslides and groundwater flooding in vulnerability assessments is limited [36]. Fatorić and Seekamp [28] identified 16 barriers to adaptation of cultural heritage to climate change in the Southeast of the US, which they grouped into three main categories: institutional barriers (e.g., lack of political commitment), technical barriers (e.g., lack of technical expertise), and financial barriers (e.g., lack of funding). The barriers to adaptation identified by Carmichael et al. [29] at cultural heritage sites in Australia were related to governance and compatibility with current management frameworks. Casey [30] identified two categories of barriers to adapt cultural heritage sites in three US National Parks to climate change: institutional barriers, which result from existing structures and frameworks, and conceptual barriers, such as problems in prioritizing the cultural resources to be preserved and adapted, together with challenges in managing those resources. Casey [30] also identified potential solutions to overcome those barriers, for example, including climate change in regulations and management plans.

The issue of adaptation to climate change began to be seriously considered only since the publication of the UNFCCC in the 1990s [11,37]. Climate change adaptation is thus a relatively new challenge and this is particularly the case in the field of cultural heritage. The impacts of climate change on cultural heritage were first mentioned in the chapter on Europe of the Fifth Assessment Report (AR5) of the Intergovernmental Panel on Climate Change (IPCC) in 2014. At the 21st session of the Conference of the Parties (COP21) of the UNFCCC in Paris, a number roundtable discussions on the topic of climate change and cultural heritage were organized [38], and, following the adoption of the Paris Agreement, the World Heritage Committee of UNESCO, at their annual session in 2017 in Krakow, Poland, noted the pressing issue of climate change impacts on World Heritage properties and requested that the UNESCO World Heritage Centre support State Parties in managing climate change impacts and in strengthening collaboration with the UNFCCC and the IPCC [39]. The World Heritage Committee is responsible for the World Heritage Convention and consists of representatives from 21 of the State Parties, which are elected by the General Assembly. This led to the IPCC consulting with UNESCO to identify topics to be incorporated as part of AR6 to be published in 2021. Climate change impacts were only briefly mentioned in AR5 and the impacts presented in the report were mainly the results of initiatives funded by the European Commission (EC) over the last 15 years such as the Noah's Ark (2003–2007) and Climate for Culture (2009–2015) projects. These projects focused on the threats of climate change on cultural heritage with limited attention given to adaptation of cultural heritage to climate change.

There are only a few well-documented cases of adaptation of cultural heritage to climate change with most research focusing on the identification of the risks of climate change and the provision of guidelines and suggestions to adapt to those risks. A limited number of studies have examined the barriers to adapt cultural heritage sites to climate change, albeit the literature on this topic is not yet comprehensive [40]. Heritage managers need to mitigate against the impacts of current changes in climate, including changes in the frequency and intensity of extreme events [1], but also require awareness of the potential impacts of climate change. For this reason, it is recommended to include climate change within management plans and decision-making strategies [15,26]. A number of questions remain insufficiently addressed in the literature, notably, how do experts involved in the preservation of cultural heritage consider climate change? Is climate change currently included in the management of heritage sites and, if so, how? What are the opportunities and limitations in adapting cultural heritage to climate change? The answers to these questions are essential for the development of adaptation measures and to identify future research directions.

3. Materials and Methods

Semi-structured interviews were conducted in three European countries: the UK, Italy and Norway. The interviews centered on the following four questions: (1) is adaptation of cultural heritage to climate change possible? (2) Are you aware of any example(s) of management methodologies to mitigate the impacts of climate change on cultural heritage? (3) Are you aware of best practice examples in adapting cultural heritage to climate change? (4) What are the determinant factors for implementing adaptation of cultural heritage to climate change? A total of 45 interviews were conducted with experts involved in the preservation of cultural heritage. The interviewees were categorized into three main groups: 19 academics and researchers from universities and research centers in the UK, Italy, and Norway, including investigators having worked or currently working on EU projects focusing on the theme of climate change and cultural heritage; 12 members of governmental institutions working on the preservation of cultural heritage; and 14 people involved with the management of UNESCO WHS, such as managers, coordinators, and professionals. The interviewed experts on cultural heritage preservation have diverse backgrounds and specializations, including anthropologists, archaeologists, architects, conservation scientists, geologists, biologists, heritage site managers and coordinators, sustainability officers, and urban planners. The interviews were audio recorded and subsequently transcribed and analyzed using the NVivo software. (Version 11, QSR International (UK) Limited, Daresbury, Cheshire, UK).

Since there is limited research on adaptation of cultural heritage to climate change, the adoption of a qualitative methodology using semi-structured interviews was considered the most appropriate. Michalski and Bearman [41] noted the advantage of semi-structured interviews when there is inadequate knowledge on a particular topic. The interviews were complemented with a critical analysis of the grey literature such as reports and documents gathered from the interviewees, websites of the institutions employing them or that they recommended, and books and journal articles. The former included existing management plans, national policies and guidelines, regional documents and brochures. A one-day workshop titled: "Vulnerability and adaptation of cultural heritage to climate change. Supporting decision-making in adapting cultural heritage sites to climate change" was also organized to gather further information on experts' perceptions and also to allow for the exchange of information amongst participants. The workshop took place on 22 March 2018 at the New Lanark UNESCO WHS and was attended by 22 experts in cultural heritage preservation, including participants from academia, government institutions, and managers of heritage sites. The workshop included a site visit, a number of presentations, and an interactive part. The latter consisted of three facilitated roundtable discussions on issues related to the vulnerability and adaptation of different WHS to climate change. With regard to adaptation, the discussion particularly focused on the challenges of adaptation and opportunities for the implementation of adaptation strategies. The methodological approach selected for this study, which involved stakeholders' interviewees and a participatory workshop, has previously been used in climate change adaptation research [42,43]. Ethical approval was sought and obtained through the University of the West of Scotland procedure.

4. Results

The first part of the investigation focused on understanding the participants' perceptions on the possibility of adapting cultural heritage to climate change. How do they feel about it? Is adaptation of cultural heritage to a changing climate a challenge that can be dealt with? What are their doubts and perplexities on this topic? The majority of the participants agreed that adaptation of cultural heritage to climate change is possible (Table 1). Some interviewees were not fully confident about this possibility and a few did not answer this question. No interviewees indicated that adaptation is not possible. The interviewees' answers on the possibility of adapting cultural heritage to climate change were coded into three main categories to highlight their perceptions on the 'opportunities', 'requirements', and 'barriers' in adapting cultural heritage to climate change. Table 2 summarizes those 'opportunities', as identified by the interviewees, which were classified according to the following themes: 'moving from reactive to proactive adaptation', 'working on mitigation and adaptation together', 'strengthening monitoring and maintenance', 'making adaptive change', 'increase collaboration', and 'positivity'. The 'requirements' for more information and resources to make adaptation possible, as expressed by the interviewees, are included in Table 3. The interviewees also identified a number of barriers to adaptation, which were classified into the following themes: 'diversification', 'uncertainty', 'resignation', 'loss', 'preserving values, integrity and authenticity', and 'financial resources' (Table 4).

Table 1. Answer of the participants to the question "Can cultural heritage be adapted to future climate change?".

Yes	No	Yes and No	No Answer
42%	0%	31%	27%

Table 2. Opportunities for adapting cultural heritage to climate change.

Opportunities	Interviewee	Quotation
From reactive to proactive adaptation	Academic	*"Yes, it can be adapted. Considering the long-term perspective (. . .) As the climate becomes wetter with more storms it is going to be even more work to maintain the heritage as it is and increase (. . .) resilience. (. . .) How to move from reacting to small problems to consider long-term perspectives? Adapting in (the) long perspective involves choices."*
	Researcher	*"We need to adapt our thinking to climate change. We need to think in longer terms than five administration years. The problem is thinking that everything can be managed in the short term."*
Mitigation and adaptation	Academic	*"A balance (is needed) between mitigation and adaptation. We must balance both."*
	Academic	*"I think that there is too much emphasis on adaptation instead of mitigation."*
	Academic	*"With climate change the focus has been more on responding, on one hand there is adaptation, but there is more on mitigation, and trying to make the buildings more energy efficient."*
Monitoring and maintenance	Academic	*"There is a need to monitor change in cultural heritage. Monitoring is very important, and to strengthen maintenance."*
	Researcher	*"Maintenance is fundamental & necessary. Restoration is the last thing to be done. First you examine. Then therapy, and last you perform surgery. You do not wait until it falls down before intervening. Study more situations, do maintenance once the problem is clear and involve more expertise. There are no experts about this in the responsible authorities."*
	Manager of heritage site	*"Most traditional buildings are built in a very sensible way, they are built with natural materials, (they can be adapted) if they are maintained."*
	Conservation scientist	*"The best that we can do is to keep up proper maintenance to deal with increased rainfall."*
Make 'adaptive' change	Member of government institution	*"Yes, I think so. (. . .) Something has to change. In historical buildings that cannot handle the increase in rainfall realistic change should be possible, so they can still be used."*
	Manager of WHS	*"We have to think . . . how are we going to take actions? That may be things like recognising that we need to improve the drainage system. Potentially changing the detailing on the roofs. I think that will probably sensible if people now would change the size of the gutters, when the opportunity arises. There are small things that will change the appearance of the building but I would say that look after it is better. I think there is a kind of better maintenance looking at what all kind of changes would be."*
	Academic	*"With climate change you are expected to see various changes taking place. (. . .) Buildings have to be adapted to meet the environmental changes (. . .) we may try to keep the same look but you will not be able to just leave it like it was. (. . .) But on the other hand you can still retain the basic elements of the buildings even if there are changes. You can still have the facade (. . .) but you cannot pretend that there has not been change when it is been adapted."*
	Member of government institution	*"In terms of climate change adaptation and resilience it is need to be changes to that design in terms of gutters and mortars."*
Increase collaboration	Academic	*"If experts work together maybe it is possible to find solutions. Collaboration is fundamental. Nowadays the sector is abandoned to itself. Points of view must be shared."*
	Manager of WHS	*"We have close collaboration with other organizations that are very beneficial for everyone on a wider level."*
	Manager of WHS	*"I think that the international community need to work together. To raise the collaboration between countries and the level of each country. If a country has more welfare than another, it is easier to do something. (. . .) (like) sharing resources and knowledge."*
	Manager of WHS	*"The most positive impact I could imagine from climate change is if people are forced to work together."*
	Researcher	*"Involving more expertise, it would be essential to have a team . . . "*
	Academic	*"(. . .) participation between different sectors in contributing to the safeguarding of heritage and (. . .) communication and collaboration between different sectors: academic and political."*
Positivity	Member of government institution	*"There is already this sensibility (to climate change or to adaptation) in some institutions. I am sure that it can be and must be done and that there is the sensibility to do it."*
	Academic	*"I think it is possible if we work together."*
	Member of governmental institution	*"I think generally yes. (. . .) I think that it is possible as long as you do the right things."*

Table 3. Requirements for adapting cultural heritage to climate change.

Requirements	Interviewee	Quotation
Information on values, integrity and authenticity	Academic	"How many changes can you make? How much adaptation can be done to maintain the integrity? I think it has to be incorporated into management plans and it is not incorporated right now."
	Academic	"You have to do, in some cases, more dramatic changes in the building even if it is cultural heritage but this depends about the values that cultural heritage has."
Resources	Manager of heritage site	"I hope so. It is just a matter of resources. If you do not have resources you do nothing. Resources must be given but you need a national plan that foresees a self-sufficiency and a return on these interventions."
	Manager of heritage site	"Yes, we certainly can. If there are resources made available to help the plans for the future. (. . .) We need to articulate what the problems are (. . .) and have more funding for research and for making people work in this area."

Table 4. Challenges in adapting cultural heritage to climate change.

Barriers	Interviewee	Quotation
Diversification	Academic	"I think that the situation is too diversified that it is not possible to speak about such a big group of cultural heritage"
	Biologist	"Well, it depends, on the specific geographic contexts"
	Anthropologist	"I guess it would depend on the type of change happening here. It probably depends on the site. (. . .) If it is coastal there are problems."
	Academic	"I think that there is some cultural heritage that can be adapted and some that cannot be adapted. An example is a structure failing on a coastline that is eroding. In some cases it is not possible to move it. Also, in the Arctic where the permafrost is melting."
Uncertainty	Coordinator of heritage site	"It depends on how climate change will be. It is a matter of when we will face the situation. There is not a 100% certainty . . . "
	Sustainability officer of heritage site	"The problem with adaptation is that we have to cope with problems that are not visible really often."
	Academic	"Climate change has not been quantified so I would not be able to tell exactly that this area needing more attention."
Resignation	Sustainability officer of heritage site	"When we look at different types of assessment, every site is different. (. . .) For (. . .) most of the places apart from the coastal areas it is actually possible (. . .) (but) (. . .) if we continue to do as we do now we will arrive to a point that is not possible to adapt anymore."
	Manager of heritage site	"Lots of these building were original stone works but they were coated in lime. (. . .) The reason why we don't do that these days is because we are too lazy to maintain it. (. . .) If we recoat the lime mortar, we will get dirty every couple of years so we have to paint it . . . "
	Conservator scientist	"Cultural heritage doesn't adapt itself. Is the things that we can do for trying to protect it from rainfall. (. . .) There is not so much that we can do. About roofs and gutters we can improve that situation. But with ruins of buildings there is not so much that we can do for them. There is not so much that we can do for rising sea level. If you get increased coastal erosion there are few places where you can make interventions like protective sea walls and barriers . . . "
Loss	Member of governmental institution	"I think that something will be lost. Definitely. And we do not have the resources to adapt everything. I think there are difficult decisions to be made."
	Academic	"For sure . . . for the most appreciated monuments you can do something. But our cultural heritage it is so modest; it is timber, which is really vulnerable to biodegradation and dependent on climatic conditions. I am not sure on how much we can do with the big majority of timber houses. I think that we need to accept the loss. Climate change will speed up the decay of cultural heritage."
	Conservator scientist	"A lot of coastal sites are just going to be lost, because we do not have the resources to protect them."

Table 4. *Cont.*

Barriers	Interviewee	Quotation
Values, integrity and authenticity	Academic	*"As a functional building yes. But if you take in consideration the history and the authenticity … "*
	Academic	*"You will need to make changes to it as well. You will not be able to preserve it exactly how it is because of increasing requirements."*
	Member of governmental institution	*"There are certainly a couple of Neolithic sites … that have been moved. So imagine (an archaeological site), what do you think that this is from a philosophical point of view? You made a new fake (site)."*
	Academic	*"You cannot build all coastline defences. There is not just the buildings there is also the natural landscape. There are no easy answers."*
	Manager of WHS	*"I think that we need to be careful. (…) If you change the natural balance of something you can have other effects, so you have to be really careful when you do interventions like that. Also you never want to lose the original fabric."*
Financial resources	Academic	*"Depends on the scale of event. If we think about the coastal erosion in archaeological sites, there are things that we can try to do, like build defences. But they will be very costly. We record with laser scanner technology. We move the thing? It is very difficult."*
	Conservation scientist	*"There is a lot of coastline and vulnerable sites, not just properties that are in care, but archaeological sites, where there isn't money or resources to protect all of them."*

The second part of the investigation focused on the interviewees' awareness of existing assessment frameworks for adaptation and best practice to preserving cultural heritage from the threats of climate change. The majority of the interviewees were not aware of strategies considering climate change adaptation in the management of cultural heritage assets and sites (Table 5). The few strategies that were mentioned are to the work done by UNESCO, Historic Environment Scotland (HES), and the guidelines of the Italian national strategic agenda (Table 6), as Italy and France are two countries that currently include cultural heritage in their National Adaptation Plan to climate change [38]. HES is the public body for the investigation, care, and promotion of Scotland's historic environment.

Table 5. Answer of the interviewees to the question "Are you aware of management methodologies to preserve cultural heritage from the implication of climate change?".

No	Some Examples *	No Answer
60%	15.5%	24.5%

* See Table 6.

Table 6. Management methodologies known by the interviewees for the incorporation of adaptation to climate change in the preservation of cultural heritage.

Interviewee	Quotation
Sustainability officer of heritage site	*"I read the one released by UNESCO; (it is) more specific for a natural site more than urban site. It is a really nice methodology because it talks about identifying the vulnerabilities, identifying risks but also identifying what the values are that you have to preserve. That it is a really nice guideline. I think that no one is using it. I haven't seen any report based on that assessment, but maybe there are and I haven't seen them. This is the only one that I am aware of. It comes to UNESCO sites and management for adaptation strategies."*
Academic	*"I am aware of a lot of work that (a national heritage authority) did which looks advanced, they have done a lot of work on the understanding of how do we manage the cultural heritage and physical assets under climate change, more rainfall, more wind and so on."*
Member of governmental institution	*"We have corporate plans about climate change. (…) The action plan identifies the specific actions on how the heritage manager will look at these things. I think that we are trying to build a management structure."*
Member of governmental institution	*"Within (our organisation) it has been an objective to try to address the impacts of climate change and then specify how we are going to do that."*
Member of governmental institution	*"We have got a number of systems here within this organization. It used to be the monument management system where they are looking at the buildings in a formalized list of things every year/two years. And there is a new monument management system coming out at the moment, which takes into consideration climate change, includes visitor access, visitor safety, and general wide things about the monument. And I know that we have the risk assessment."*
Academic	*"Well, there is the 'agenda strategica nazionale italiana' (Italian national strategic agenda)."*
Academic	*"A little bit. I am aware of the work done by (a national heritage authority) particularly with climate change impact for Orkney. Also about the giant causeway in Northern Ireland. It is a natural heritage site, but it suffering from sea level rise. They are working with documentation. Including continuously monitoring of the site."*

The interviewees were aware of examples of best practice in adapting cultural heritage to climate change risks. Their answers were divided into 'managerial and decisional adaptation' and 'practical adaptation' following on the classification previously used in the literature [1]. The best practice examples in managerial and decisional adaptation to climate change were as follows: to increase fundraising, increase the production of knowledge and its dissemination, engage those involved with the heritage (owners, communities, tourists) in adaptation, promote and strengthen monitoring and maintenance, upgrade management plans to include climate change, strengthen regulations and guidelines, keep working on mitigating climate change to reduce future risks (Table 7). Examples of practical adaptation to climate change identified by the interviewees included building defences, using roofs and shelters to protect unroofed sites, upgrading roofs and drainage systems, avoiding the use of incompatible repair materials and surface treatments, moving the heritage sites, monitoring the heritage assets and the climatic conditions, and the use of digital recording (Table 8).

Table 7. Best practice in climate change adaptation identified by the interviewees: managerial and decisional adaptation.

	Managerial and Decisional Adaptation
Financial	*"Funding programs and grants (to help) owners."*
Financial and knowledge	*"Being more aware of what the problems areas are. Just target funding. Knowing that when you do repairs you need to know the stone specification and the right mortar mix, the lead detail installed appropriately to high standards. So it is really attention to details. Do the work for repairing like for like. (. . .) So awareness of these (technical material) things; for that you need qualified professionals to be involved. Conservation accredited professionals and they are working with these schemes they are more likely to get a good result."*
Knowledge	*"Improve the quality of information."*
	"Helping owners of properties understand the implications of what is coming."
	"Good understanding of what the threats are. (. . .) Having options to choose on how to adapt."
	"Improve the quality of the information."
Knowledge and dissemination	*"Better disseminate the already present solutions."*
Knowledge and engagement	*"Demonstrating the works. (. . .) Demonstrating very practical tools and activities where people can see the work."*
	"Education."
	"Work for better communication between different sectors (. . .) for the engagement of the community (. . .) for the increase of knowledge."
Engagement	*"Involving communities, making them more aware, engaging them. (. . .) I think that this actually has to come from the government and from the local communities. They have to say we have to do something about it."*
	"Encourage people to adapt."
Monitoring and maintenance	*"Very good maintenance."*
	"Promoting positive maintenance and repair on historical buildings."
	"Monitoring to have a starting point to understand the problems on the sites and improving the management of heritage."
	"Monitor where the impacts are more severe."
	"Very good maintenance."
	"Extend the maintenance not only to the objects, but also to the surroundings."
	"Coastal protection, regular maintenance . . . "
Management	*"Make changes in plans."*
	"Include climate change in management plans."
	"Improve the management at local level."
Value preservation	*"We do not want to lose the original appearance. We do not want to people lose the original design just because they have to adapt. We have to be clever on how we do it. Same shape and materials, keeping good records of what there was before, so people have the knowledge."*
Regulations	*"Having more rigid protocols."*
	"Local government needs to give guidelines."
Mitigation	*"Work also on mitigation. Unless for some aspects there is a point of no return we still need to work on mitigation."*

Table 8. Best practice in climate change adaptation identified by the interviewees: practical adaptation.

	Practical Adaptation
Building defences	*"Some sites like Skara Brae there is a sea wall in place for a long time to reduce the amount of coastal erosion."*
	"Build defences where natural hazards can (impact)."
	"Flood defences. Venice is a good example."
Roofs and shelters	*"In some archaeological sites they built a roof, a tent, with modern materials on them. I think that there are challenges in how to do it considering the broader landscape as well."*
Roofs and drainage	*"Avoiding water penetration in the structures. Maintaining roofs and gutters and other protective layers. Keeping the structures dry."*
	"Methods for dealing with water shedding for buildings. If you have heavier individual rainfall events, could be useful to increase the capacity of the building to shed water quickly. You can make sure that there is a very good maintenance of the roof, of gutters and downpipes to get rid of water. I am aware that in some cases with increased capacity of gutters and downpipes . . . They can't be overwhelmed by the volume of water, otherwise it overflows and can come into the building."
Drainage	*"Deal with damp penetration and moisture post flood. Increased drainage."*
	"Adapting the gutters."
	"Enlarging the drainage to cope with the increased rainwater."
Move	*"Move the heritage. In Norway they are already doing it. The wooden houses close to the coast that are at risk of coastal related issues have been moved in the (inner) land."*
Building materials: Avoid cement mortar	*"There are some structures that have particular vulnerabilities to increased rainfall (. . .) built with impermeable stone like granite and (. . .) with cement mortar, and the combination of the impermeable cement mortar and impermeable stone is (. . .) very bad (. . .) when there is (. . .) especially (. . .) heavy rainfall. Because there is no capacity in the building to deal with penetrating water that tends to come straight to the walls, those types of structure are very vulnerable to increased rainfall. You can't protect them completely (. . .) if you had infinite capacity would be to rebuild them without the cement, because if you use lime mortar, then the wall has more capacity to absorb the rainfall like a sponge. It absorbs the rainfall and then evaporates later. But with all the cement it comes straight through. But of course you can't take down all these buildings and rebuild them. In some sites, there are possibilities to do repointing, with a lime mortar that has more capacity to deal with rainfall that increases the resilience."*
Building materials: Harling of surfaces	*"(. . .) harling of surfaces, when you have a stone surface you can cover it with a lime mortar. And quite a lot of buildings originally have had a lime harling on them, it gives the surface the capacity to deal with rainfall and absorb it, rather than letting the walls absorb it and hold it, and when it stops raining evaporates. That can increase the resilience of the properties. Some buildings originally had harling that has been (. . .) replaced. (. . .) Tends to be quite controversial to do that because people tend to be used to the appearance of stones. And people like the appearance of stones (. . .) (harling) changes the appearance (of the buildings) and people can be quite resistant to the idea of doing that."*
Buildings materials: Avoid moisture barriers	*"These kind of water repellents (. . .) tend to be a disaster. Because water always find a way to come inside. If there are cracks inside the wall. At the bottom of the wall, water can be dragged on by capillary action. If it is an inhabited building then you have internal humidity and if you use water repellent barriers on the inside and it is not properly ventilated it keeps the water inside. So there are always in historical buildings (. . .) this bad situation. A modern structure tend to be designed with the use of barriers. (. . .) But historic buildings cannot work with barriers, because it keeps the problems much worse."*
Building materials: New materials	*"Work for developing new materials that can protect the historical ones. That can be reversible. Test them on case studies first."*
Monitoring	*"Keeping regular checks of the conditions of the monuments. (. . .) regular monitoring (. . .) make sure that the stonework is not being destabilised by the loss of mortar and by the decay of the stone."*
	"Put sensors for the monitoring of temperature, humidity, for estimating a loss of colour in the artefacts, etc. that needs to be used effectively for the intervention in case of an event. The monitoring has to be "usable" and understandable. (. . .) In our site we developed a system for the monitoring of condensation phenomena on frescoes linked with a retro-heating system that keeps the temperature of the frescos over the condensation point."
Digital recording	*"Perpetuate the memory of people's life (. . .) today there is the digital. (. . .) digitalise it and make it visible to everyone."*
	"Recording, use high tech digital recording. Scanning."

The interviewees were then asked to identify the determinant, or constraining, factors that they consider important for adapting cultural heritage sites to climate change. A list of these factors is provided in Table 9 and categorised into the following groups: (1) Knowledge, education, communication, and awareness; (2) Management, regulations, governance, and drivers; (3) Economic factors; (4) Cultural values; (5) Health and safety concerns; (6) Time.

Table 9. Determinant factors for adapting cultural heritage to climate change as identified by the interviewees.

Determinant Factor for Adapting Cultural Heritage to Climate Change:	
Knowledge, education, communication and awareness	*"Knowledge"* *"Lack of example to follow"* *"Lack of traditional skills"* *"Need for better workforce, need for people trained in traditional skills"* *"Knowledge about the traditions and the traditional skills"* *"Procurement of traditional materials"* *"Proper conservation (knowledge) available for the final users"* *"Technologies"* *"Lack of definitions"* *"Education"* *"Community and owners awareness and engagement"* *"Make sure that all the stakeholders involved are clear about the co-benefits"* *"Knowledge about the history to make people proud"* *"Cultural importance of the buildings"* *"Information"* *"Awareness"* *"Communication"* *"Increasing awareness and education in young age"* *"Changing behaviour. Prioritize maintenance and repairs of gutters versus a new kitchen, for example"* *"Mainstreaming"* *"Having knowledge about the risk you can cope with"* *"Be aware of the problems"* *"Having more knowledge on the possible scenarios"* *"Translating scientific research into something usable for the stakeholders involved in the maintenance and repair"* *"Continue monitoring"* *"Having knowledge on how to prepare the plans, maintain the buildings, prepare them for risk, strengthen the structure and materials" "Manager with technical knowledge"*
Management, regulations, governance and drivers	*"Need for driver to push people working in government"* *"Force people legally to do something"* *"A broad regulatory picture"* *"Upgrading building regulation to apply to existing buildings"* *"Stronger regulation"* *"Capacity to get things done. Taking decisions"* *"Strong advocacy for the management itself. The management have to protect the authenticity"* *"Make sure that climate change is part of the managing plan"* *"Clear laws"* *"Having management plans related to laws"* *"Have politician that understand the issue"* *"Cooperation with the communities"* *"Cooperation between different levels (e.g., management, local authorities, governmental institutions, etc.)"* *"Development of good plans"* *"Push from government"* *"Collaboration with governmental institutions or with the government"* *"Ability of organization"* *"Responsibility"* *"Willingness of national authorities"* *"Rules"* *"Standards to follow"* *"Trained staff"* *"Legislations at political level"* *"National legislation"* *"European legislation"*
Economic factors	*"Financial limits"* *"Incentives"* *"Funding programs"* *"Financial support from government"*
Cultural values	*"Cultural heritage values"* *"Different type of values"* *"Acceptance that adaptation is necessary"* *"Philosophical constraints"* *"Cultural constraints"* *"Cultural awareness"*
Health and safety concern	*"Health and safety concerns"*
Time	*"Time"*

5. Discussion

The perceptions of experts working in the field of cultural heritage on adaptation to climate change as gathered through semi-structured interviews and a participatory workshop revealed a number of results that are in agreement with the literature, however, issues not previously reported emerged from this study, notably on the opportunities for, and barriers to, adaptation.

5.1. Opportunities for Adapting Cultural Heritage to Climate Change

5.1.1. Moving from Reactive to Proactive and Planned Adaptation

The interviewees expressed the need to move from reactive to proactive adaptation. Reactive adaptation is in response to the impact of a climatic event that has already been experienced, for instance a severe storm, landslide, or flood, and often results in short-term planning that aims at returning to conditions as they were prior to the event in question without necessarily aiming to increase the resilience of the site to future events. Proactive or anticipatory adaptation takes place before the occurrence of a climatic event and can be in response to projected changes in climate [44], thereby allowing more time for consultation, the discussion of alternatives, and long-term planning. Such a proactive, planned adaptation involves a policy decision (for cultural heritage this can be a managerial decision) for adapting cultural heritage to changing climatic conditions. During the interactive session of the workshop, some of the participants shared the opinion that moving from reactive to proactive adaptation is feasible: *"Long term proactive planning should be compatible with existing mechanisms for ensuring management of heritage sites. It should be possible to factor climate change into existing processes."* (Workshop participant, member of a governmental institution). There is still time to *"assess, understand and forecast the effects of climate change on cultural heritage in decades to come, before effects are seen, from which relevant mitigation and protection of sites can be developed to prevent a loss of heritage"* (Workshop participant, academic). This, however, also constitutes a challenge: *"I think there are challenges in getting heritage and other asset managers to take ownership and action on this issue."* (Workshop participant, member of governmental institution). Moving from reactive to proactive adaptation is essential to increase resilience at cultural heritage sites to future changes in climatic conditions, including extreme weather events. Once knowing the temporal probability of hazard scenarios, it can be done, for example, beginning with an assessment of the vulnerability of heritage assets or a specific site to understand baseline conditions, monitoring the condition of the site over time frames comparable with ones used by climate scientists in order to minimise natural weather variations, operating continuous maintenance, prioritizing the most vulnerable elements, and planning adaptation options according to projected changes in climate.

5.1.2. Working on Mitigation and Adaptation Together

The interviewees indicated the need to pursue efforts aimed at reducing GHG emissions to reduce future risks of climate change on cultural heritage in addition to adapt to the existing and unavoidable risks of a changing climate. There are a number of mitigation actions that can be undertaken in the field of cultural heritage, for example, enhancing energy efficiency in historical buildings, improving the sustainability of interventions in historical buildings, enhancing the reuse of original materials during restoration and refurbishment, and promoting sustainable tourism. Both adaptation of cultural heritage to climate change risks and mitigating climate change to reduce the carbon footprint of heritage buildings are needed and complementary, as well as being recommended by UNESCO [45].

5.1.3. Strengthening Monitoring and Maintenance

Many interviewees found that strengthening monitoring and maintenance of cultural heritage is necessary to cope with climate change impacts. The monitoring of climatic conditions and of changes in the condition of heritage assets through surveys, for instance, also creates a baseline dataset against which the impacts of projected changes in climate can be estimated. If a change in decay is predicted as a result of a vulnerability assessment, for instance, it might be possible to manage and mitigate this risk through appropriate maintenance, thereby minimizing or avoiding the damage that would otherwise occur and the need for expensive restoration work. This can be done, for example, by simulating scenarios of possible change in decay with regular time frames, for example every five years, using decay functions and models that use data on material properties, the geometry of the objects and climatic conditions. Such views from cultural heritage experts are consistent with the guidelines from UNESCO and ICOMOS, which suggest that long-term programs of monitoring and maintenance should be implemented prior to applying any adaptive measures [19,45]. This is also supported by Sabbioni et al. [1] who recommended more frequent monitoring and maintenance of cultural heritage sites.

5.1.4. Making Adaptive Change

There is a dichotomy in the interviewees' perceptions with regard to making adaptive change to cultural heritage. Some see 'change' resulting from adapting cultural heritage to climate change as an opportunity (Table 2) and others perceive it as a challenge (Table 4). When you make changes to cultural heritage there is always the possibility of a conflict between the actions associated with the change and the protected values, integrity, and authenticity of the heritage. A suggestion given by some workshop participants was the development of tools to visualize this 'change' prior to adopting it as a way to predict and argue on the potential decrease in the significance of the cultural and natural heritage (the two are often interconnected) of the protected site. Moreover, such an approach might also avoid the selection of an unsuitable, or not properly studied adaptive solution as mentioned by an interviewee when giving the examples of constructing sea barriers to protect cultural heritage sites located in coastal areas at risk. In Venice, for example, the mobile gates of the *MOdulo Sperimentale Elettromeccanico* (MOSE)—Italian for Experimental Electromechanical Module—defend the city from sea level rise, high tides, and flooding, and the outcome of such an adaptive solution is overall positive. There are other situations, however, where the selected adaptive solution is inappropriate, for instance, the breakwater barriers built in Agropoli in the Italian province of Salerno led to the proliferation of the *Posidonia oceanica* algae, which has negative consequences on the marine biodiversity of the area. *"It should be known in a better way these technological solutions ... How climate change can have repercussion on the territory?"*(Interviewee, researcher). Adopting wrong solutions can lead to a reduction of the multiple values associated with cultural heritage. For instance, decreasing the aesthetic and architectural value of a cultural heritage resource can decrease the interest of tourists with a subsequent decrease in economic and social values or, equally, leading to disinterest of local communities from protecting the heritage site. One could argue that 'adaptive change' can be considered an opportunity only if there is enough research behind the solutions to be adopted both in terms of preservation of heritage values and in terms of the verification of the performance and efficiency of the selected solutions in the long-term.

5.1.5. Increase Collaboration

The interviewees emphasized that increased collaboration between different experts, disciplines, institutions and countries presents an opportunity to better implement adaptation strategies. A few interviewees mentioned the lack of communication between different sectors (e.g., academia, government institutions, and local management). Some of the site managers interviewed were not aware of relevant published academic research or of guidelines published in national documents and UNESCO reports. This also highlights a lack of dissemination of the outcomes of academic research and initiatives from government institutions. Collaboration between different actors such as governmental bodies, private individuals, and communities was defined both as a need and as a barrier, the latter when it was non-existent, in the US [28]. International collaboration and cooperation between heritage organizations is promoted by the EC, UNESCO, and ICOMOS [45,46].

5.1.6. Positivity

Some interviewees are confident of the possibility to adapt cultural heritage to climate change impacts. The willingness to move forward adaptation of cultural heritage to climate change is a stimulus towards proactive management and hence a determinant factor for the change itself. If people are positive about it, they might invest more resources in proactive adaptation and research.

5.2. Requirements for Adapting Cultural Heritage to Climate Change

5.2.1. Information on Values, Integrity and Authenticity

Some interviewees mentioned the need for more information and guidelines on adaptation of cultural heritage to climate change in relation to protecting the values, integrity, and authenticity of cultural heritage. There was a perceived need to investigate further available options to preserve cultural heritage in the context of climate change. Adaptation will possibly be limited if the safeguarding of authenticity, integrity, and value of cultural heritage is considered. *"How many changes can you make? How much adaptation can be done to maintain the integrity?"* (Interviewee, academic). The problem of altering the authenticity of heritage while restoring it is a perennial problem that cultural heritage faces. This is made difficult by the paucity of practitioners with traditional construction and conservation skills due, in part, to a lack of such training opportunities to architecture and engineering students [47,48]. Previous research stresses the need for more technical information and better research on this topic [27,28]. Phillips [27] emphasized the difficulty of finding resources and guidance on climate change adaptation and highlighted that the current tools available from national authorities and governmental institutions for climate change adaptation are not used by experts working in cultural heritage.

5.2.2. Resources

At the stakeholders' engagement activities, it was noted that more resources (financial, technical, and human) are required to enable adaptation of cultural heritage sites to climate change. More research (and funding for research) was also perceived as a need to inform the adaptation process. Investigating climate change impacts at cultural heritage sites, the monitoring and maintenance of the sites and the implementation of adaptation measures require money, knowledge, skilled people and new technologies. Della Torre [47] stated that adapting means thinking about how to improve the management of heritage sites, not only for monitoring and maintenance but also fundraising, marketing, and protection in view of climate change impacts.

5.3. Barriers in Adapting Cultural Heritage to Climate Change

5.3.1. Diversification

The interviewees highlighted the difficulty in generalizing adaptation solutions due to the diversity of typologies of cultural heritage, the different geographical locations of heritage assets and the context in which they are located, and the climatic conditions to which they are exposed. This is in addition to the state of decay and the different materials, geometry, and age that characterise cultural heritage assets. As a cultural heritage site is embedded within its environment, one cannot disregard the surrounding landscape in the conservation and valorisation practices [47]. Also, because every site and/or asset has its own cultural, social, and economic values, the interviewees found it difficult to make generalisations on the adaptation practices to be adopted. This supports the view that climate change adaptation requires a case-by-case approach. Nonetheless, there are some types of adaptation practices that can be generalised, for example, strengthening monitoring and maintenance of the sites, and increasing awareness of climate change impacts. Furthermore, adaptation can happen through small steps and at different scales: national (e.g., implementation of regulations), regional (e.g., engagement of population), local (e.g., updating management plans), or done autonomously (e.g., improve regular maintenance and repairs) [19].

5.3.2. Uncertainty

A number of interviewees identified the uncertainty in the climate change projections and in the impacts of climate change as well as the effectiveness of the adaptation solutions to deal with those impacts as barriers to adaptation. Existing climate change projections were not known or used by many interviewees engaged with the management of heritage sites at the local level. Other interviewees and workshop participants stressed the difficulty in using climate change projections because of the lack of availability of a user-friendly interface and the lack of detailed information provided by the projections. In addition, the lack of information on climate change impacts and uncertainty in the effectiveness of adaptation measures that can be adopted were issues of concern, especially for highly vulnerable sites and where there are limited options for adaptation. The workshop attendees also identified the need for "*developing accurate long-term plans and forecasts of the future effects of climate change.* [...] *Understanding how different sites will react to different pressures posed by climate change*" (Workshop participant, academic). Uncertainties in climate change projections are commonly referred to as a barrier in the literature. Carmichael et al. [21] mentioned the limited availability of high-confidence climate change projections at the local scale. Phillips [27] also raised the issue of uncertainty in the climate change scenarios as well as the difficulty in interpreting them.

5.3.3. Resignation

Some stakeholders expressed despair regarding the possibility to adapt cultural heritage to climate change, because of limitations such as the lack of technologies and resources and low confidence in peoples' willingness for behavioural change, for example, laziness and lack of commitment to mitigate against climate change impacts. It is written in the literature that as long as our heritage is not lost, something can still be done with regard to climate change impacts [1,45]. Preserving our cultural heritage requires a positive and constructive attitude and brings new challenges such as users' engagement (e.g., visitors, staff at cultural heritage sites), research, and management processes, which can be dealt with by adopting cooperative approaches and the co-production of knowledge.

5.3.4. Loss

The majority of the interviewees agreed that it is probable that some our cultural heritage will be lost as a result of climate change. This is particularly the case in coastal regions affected by coastal erosion and in regions with melting permafrost. A catastrophic impact would be permanent inundation of coastal heritage cities and sites as a result of sea level rise. The risk of losing part of our heritage because of climate change is exacerbated by the lack of resources available for adaptation, notably financial, human, technical, and time, and the potential disinterest from communities on preserving sites of cultural value in their proximity. The potential loss of cultural heritage as a result of climate change has previously been cited in the literature. Sabbioni et al. [1] stated the need to accept the loss of some of our heritage. UNESCO, for its part, underlined that not all cultural heritage can be saved and that some sites might be abandoned because of the consequences of extreme events [45]. In fact, Marzeion and Levermann [49] estimated that 6% of UNESCO heritage sites will be impacted by sea level rise over the next two millennia under the projected increase in temperature. Fatorić and Seekamp [28] identified potential loss of cultural heritage as a barrier in terms of "lack of knowledge about letting it go" and in terms of the difficulty from doing so from a managerial perspective. Also, Berenfeld [50] raised the inevitable need for taking 'hard decisions' when selecting which sites to condemn and which sites to preserve.

5.3.5. Preservation of Values, Integrity and Authenticity

Strong concerns were expressed by the interviewees regarding the need to adapt to climate change while also preserving the values related to the authenticity and integrity of cultural heritage. This is because some adaptation solutions might sacrifice heritage values or affect the authenticity of the site. An example made by some interviewees was the construction of shelters and roofs over heritage sites, an adaptation solution that can disturb the unity between the heritage object and the landscape, and decrease the authenticity of the archaeological site due to the addition of this non-indigenous element. Another example is the decision to move the heritage site due to its location on an eroding coastline, for instance, which inevitably puts the heritage site out of context, thereby affecting its authenticity. Climate change impacts can affect not only the physical cultural heritage, but also the intangible heritage such as memory and history and the societies and communities associated with the tangible cultural heritage [45]. The 'lived experience of culture', the 'identity', 'belonging', and 'sense of place' together with the values, traditions, and cultural practices need to be taken into consideration to determine the acceptability of the climate change adaptation solution [51]. During the workshop the need to engage not only with decision-makers but also with the communities in understanding the heritage values was debated. Some participants mentioned that non-experts in heritage conservation do not often understand the values of a heritage site. The sharing of values of the heritage sites can be accomplished through public engagement and community education on the history and significance of heritage resources. This is consistent with Fatorić and Seekamp [28], who highlighted that experts in cultural heritage preservation from the US expressed concern on the limited understanding of potential changes in the integrity of cultural heritage as a result of climate change adaptation.

5.3.6. Financial Resources

The lack of financial resources available to cope with climate change impacts was another issue identified by the interviewees and workshop participants. In heritage sites with high vulnerabilities costly adaptation measures are required, however, institutions and managers in charge of the preservation of cultural heritage sites do not have sufficient financial resources to undertake all required adaptation efforts. It is impossible, for instance, to move all the archaeological sites at risk from coastal erosion. In Scotland only, for example, there are 11,500 coastal archaeological sites [52]. Site prioritization and fundraising strategies are needed to adapt to future climate change. During the workshop the participants stated that physical adaptation is entirely dependent on the economic situation and that more economic funding for technical research is needed. Nonetheless, it was also highlighted that the need for money does not have to be more important than other solutions like public engagement, collaboration, and communication and that decision-making should not be influenced only by the need for finance. Financial resources have also been identified as a barrier to adaptation in the related work of Fatorić and Seekamp's [28] and in Phillips [27] in the US and the UK, respectively.

5.4. Management Methodologies and Best Practice

The interviews revealed a lack of awareness of methodologies incorporating climate change adaptation in the management of cultural heritage. This suggests not only the need for further research in this area but also better dissemination of the already available scientific knowledge for its consideration in the management and preservation of cultural heritage sites. This contrasts with reasonably good awareness of best practice in adapting cultural heritage to climate change by the interviewees. In all three countries investigated, adaptation strategies at the governmental level were mentioned during the interviews. In Italy, the national strategy for climate change adaptation recommends the revision of the management plans of cultural heritage sites by introducing specific measures for adapting cultural heritage to climate change [31,32]. It also provides some adaptive solutions to climate change impacts, for example, increasing knowledge, dissemination, monitoring, and maintenance, and delivers a list of best practice in relation to physical adaptation of selected heritage materials (stone, wood and metals). In the UK, HES provides support and guidance on strategies to increase the resilience of Scotland's historic environment [53]. For example, in Curtis and Hunnisett Snow [54] best practice on how to adapt traditional buildings are outlined, such as increasing their resilience through maintenance of building fabric and improvements in the drainage system. In Norway, the Riksantikvaren (Directorate or Cultural Heritage) is committed to the investigation of climate change impacts and the dissemination of adaptation measures to preserve the country cultural heritage from the impacts of climate change [55].

5.5. Best Practice in Adapting Cultural Heritage to Climate Change

The IPCC identifies four main categories of adaptation measures: technological, behavioural, managerial, and policy. A technological measure to adapt cultural heritage to climate change includes the mobile gates protecting the city of Venice against coastal flooding [56,57]. Changes in the behaviour of tourists and inhabitants living near the protected sites is an example of behavioural adaptation while the monitoring and maintenance of the sites would be considered a managerial measure. Examples of a policy measure are regulations, guidelines, and funding [58,59]. In this paper, the interviewees' answers were divided into two groups: (1) managerial and decisional, which is consistent with the IPCC categories named 'managerial' and 'policy', and (2) practical, which encompasses the 'technological' and 'behavioural' IPCC categories.

The suggested 'managerial and decisional' adaptations included the following:

- Financial resources
- Knowledge of climate change impacts on cultural heritage
- Dissemination of information
- Engagement with stakeholders (e.g., communities and decision-makers)
- Monitoring and maintenance
- Inclusion of climate change in management plans
- Preservation of values
- Regulations and guidelines for adaptation
- Mitigation strategies

The suggested 'practical' adaptations included the following:

- Building coastal defences, roofs, and shelters
- Improving drainage system
- Moving the heritage site
- Avoiding maladaptation such as the inappropriate use of certain building materials and developing new materials compatible with the historic environment
- Improve or strengthen monitoring
- Digital recording of cultural heritage

The examples of climate change adaptations suggested during the stakeholders' consultations are generally in line with the guidelines published in the literature and by governmental organizations. However, there are a few discrepancies in some suggested practical applications in the literature and amongst our interviewees. Some interviewees identified issues with the use of impermeable surface coatings such as water repellents and consolidants on heritage buildings, for instance. This is because with increasing rainfall, walls are likely to retain more moisture and the use of impermeable coatings would reduce the capacity of the excess water to evaporate from the walls. This appears to be a controversial topic with governmental institutions from different countries proposing opposite solutions. In Italy, for example, the use of water repellents and consolidants continue to be promoted as an effective strategy [31]. In contrast, HES states that external waterproof coatings should have breathable properties and thus water repellents and consolidants are not adequate as a long-term solution because they trap moisture inside the walls, thereby causing damage to the stones and bricks typically used in traditional buildings [54]. This does not mean that this adaptive solution should also be unacceptable in Italy, but that a solution in one country could not necessarily be applied in other countries, and that any adaptive option should be contextualized, detailed, and scientifically proved prior to its adoption.

Available research on traditional conservation should be used to validate climate change adaptation options. "One modus operandi or one restrictive set of guidelines or rules are certainly not going to be sufficient to preserve this immense heritage" [19] (p. 120). The scientific community should share information, knowledge, and open dialogues to understand the best solution for adaptation. Moreover, the dissemination of information needs to be clear, concise, and easily accessible, and potentially using new technologies to reach different stakeholders. The knowledge needs to be shared not only on best practice for adapting cultural heritage to climate change, but also on the maladaptive actions that can increase further the vulnerability of cultural heritage to climate change [60].

5.6. Determinant Factors

The determinant factors that were identified for the implementation of adaptation of cultural heritage to climate change are listed in Figure 1 and could be divided into six groups: (1) Knowledge, education, communication, and awareness; (2) Management, regulations, governance, and drivers; (3) Economic factors; (4) Cultural values; (5) Health and safety concern; and (6) Time (Table 9). This list can be used when planning climate change adaptation in the field of cultural heritage.

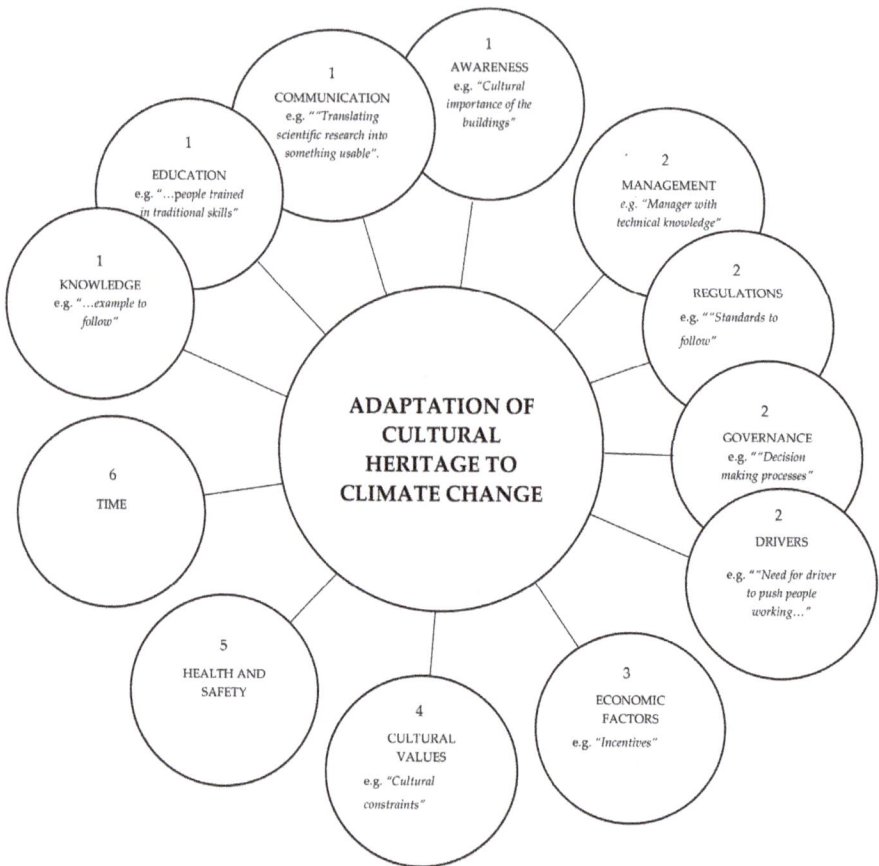

Figure 1. Determinant factors for implementing adaptation of cultural heritage to climate change, as identified by the interviewees.

6. Conclusions

Adaptation of cultural heritage to climate change is necessary to mitigate climate change impacts and to increase the resilience of historical sites. For this reason, organizations such as UNESCO, ICOMOS, IPCC, and the EC recommend more research on this topic. Accordingly, this paper analysed the perceptions of cultural heritage experts on climate change adaptation. Specifically, it reports on how climate change adaptation is considered in the management of cultural heritage sites, in scientific research and by governmental institutions and authorities working in the field of cultural heritage preservation. A common view amongst the interviewees was that adaptation of cultural heritage to climate change is possible. Opportunities for, barriers to, as well as requirements for adapting cultural heritage to climate change, as ascertained by the interviewees, provided a better understanding of what needs to be provided and prioritized for adaptation to take place and in its strategic planning. A lack of knowledge of management methodologies incorporating climate change impacts was reported albeit the interviewees were aware of a number of best practice examples in adapting cultural heritage to climate change. This paper highlights the need for more research and the identification and dissemination of practical solutions and tools for the incorporation of climate change adaptation in the preservation and management of cultural heritage. Figure 2 summarizes the research findings.

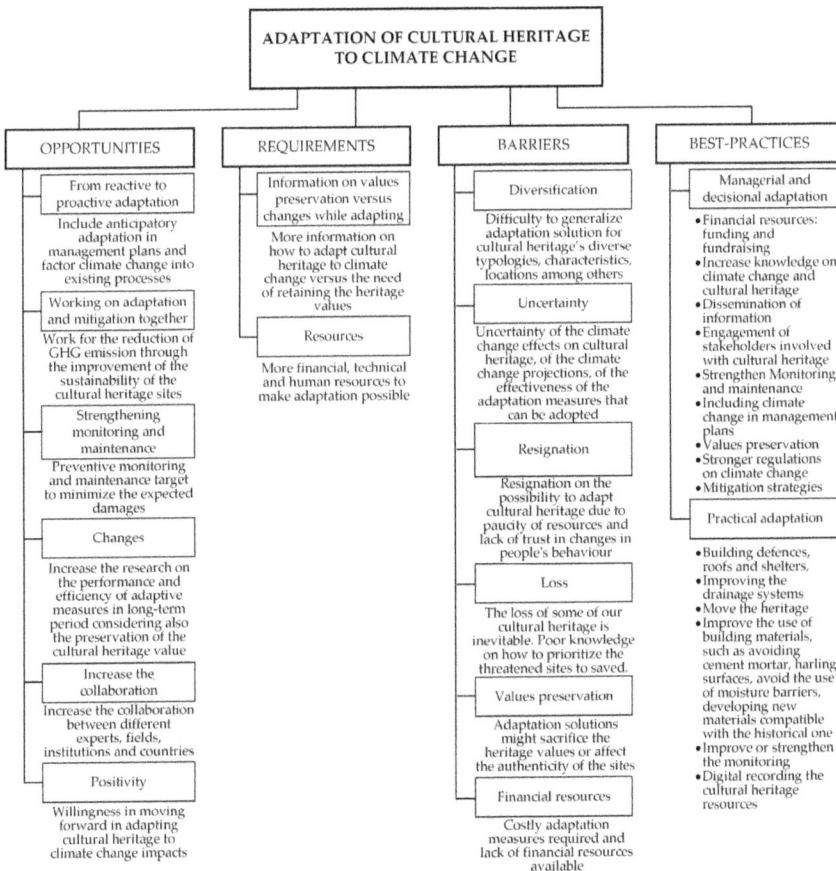

Figure 2. Conceptual diagram depicting the research findings.

This paper emphasized the complexity of the topic of adaptation of cultural heritage to climate change and the need for further research on this topic. Since the Paris Agreement, UNESCO has aimed to support State Parties in managing climate change impacts, however, State Parties should also collaborate and learn from each other. Adaptive measures developed in one country could be useful and adapted in another country, for instance. Hence the knowledge acquired on climate change adaptation should circulate freely and more developed countries, or countries with strategies, methodologies, measures, and more money for research, should help those that are less fortunate. This requires dialogue and collaboration, however, not only between national governments, but also between heritage organizations, institutions, research centers, heritage site managers, academics, and others involved in the preservation of cultural heritage. There is also an interrupted flow of information between the knowledge available at the international level and the passing of that knowledge down to the local management scale. Another recommendation is mainstreaming climate change in approaches used to manage cultural heritage sites, but this requires awareness of climate change impacts and hence greater dissemination as well as raising awareness of this issue to local communities and increase community engagement, as support from the wider community is important for successful adaptation to climate change. Regulations, the provision of guidelines, and financial incentives are also needed to commit organizations to climate change adaptation. Further research is needed on adaptive measures for specific heritage typologies or materials, mitigation of climate change in field of cultural heritage, and the managing of loss of cultural heritage as a result of inundation of coastal lands, for instance, risk-preparedness and the complexity of making adaptive changes to cultural heritage while also preserving heritage values.

Author Contributions: Conceptualization, E.S., A.G., C.B. and J.H.; Data curation, E.S.; Formal analysis, E.S.; Funding acquisition, E.S., A.G. and J.H.; Investigation, E.S., A.G. and J.H.; Methodology, E.S. and A.G.; Project administration, A.G.; Resources, J.H.; Supervision, A.G., C.B. and J.H.; Validation, C.B.; Visualization, E.S.; Writing—original draft, E.S.; Writing—review & editing, A.G., C.B. and J.H.

Funding: A university studentship was provided by the University of the West of Scotland to support the doctoral studies of the first author. The Postdoctoral and Early Career Researcher Exchanges (PECRE) funding scheme of the Scottish Funding Council and awarded by the Scottish Alliance for Geoscience, Environment and Society (SAGES) supported the academic exchanges of the first author to Italy and Norway.

Acknowledgments: We thank all the interviewees for their time and information, and Historic Environment Scotland for the grey literature provided. We are also grateful to The Institute of Atmospheric Sciences and Climate, National Research Council of Italy (ISAC-CNR) in Bologna and the Norwegian University of Science and Technology (NTNU) in Trondheim where two academic exchanges were conducted, and to the Scottish Funding Council via the Scottish Alliance for Geoscience, Environment and Society (SAGES) who funded those two academic visits.

Conflicts of Interest: The authors declare no conflict of interest.

References

1. Sabbioni, C.; Brimblecombe, P.; Cassar, M. *The Atlas of Climate Change Impact on European Cultural Heritage: Scientific Analysis and Managment Strategies*; Anthem Press: London, UK, 2010.
2. Bertolin, C.; Camuffo, D. Climate Change Impact on Movable and Immovable Cultural Heritage throughout Europe; Damage Risk Assessment, Economic Impact and Mitigation Strategies for Sustainable Preservation of Cultural Heritage in the Times of Climate Change. Available online: https://www.climateforculture.eu/index. php?inhalt=download&file=pages/user/downloads/project_results/D_05.2_final_publish.compressed.pdf (accessed on 14 August 2018).
3. Leissner, J.; Kilian, R.; Kotova, L.; Jacob, D.; Mikolajewicz, U.; Broström, T.; Ashley-Smith, J.; Schellen, H.L.; Martens, M.; van Schijndel, J.; et al. Climate for culture: Assessing the impact of climate change on the future indoor climate in historic buildings using simulations. *Herit. Sci.* **2015**, *3*, 1–15. [CrossRef]
4. Cassar, M. Climate Change and the Historic Environment. Available online: http://unfccc.int/resource/docs/convkp/conveng.pdf (accessed on 9 August 2018).
5. Brimblecombe, P.; Grossi, C.M.; Harris, I. Climate change critical to cultural heritage. In *Survival and Sustainability*; Springer: Berlin/Heidelberg, Germany, 2011; pp. 195–205.

6. United Nations Framework Convention on Climate Change. Available online: http://www.globaldialoguefoundation.org/files/ENV.2009-jun.unframeworkconventionclimate.pdf (accessed on 9 August 2018).

7. Klein, R.; Smith, J. Enhancing the capacity of developing countries to adapt to climate change: A policy relevant research agenda. In *Climate Change, Adaptive Capacity and Development*; Smith, J., Klein, R., Huq, S., Eds.; Imperial College Press: London, UK, 2003.

8. Smit, B.; Burton, I.; Klein, R.J.T.; Street, R. The science of adaptation: A framework for assessment. *Mitig. Adapt. Strateg. Glob. Chang.* **1999**, *4*, 199–213. [CrossRef]

9. Adger, W.N.; Agrawala, S.; Mirza, M.M.Q.; Conde, C.; O'Brien, K.; Pulhin, J.; Pulwarty, R.; Smit, B.; Takahashi, K. Assessment of adaptation practices, options, constraints and capacity. In *Climate Change 2007: Impacts, Adaptation and Vulnerability*; Contribution of Working Group II to the Fourth Assessment Report of the Intergovernmental Panel on Climate Change; Parry, M.L., Canziani, O.F., Palutikof, J.P., van der Linden, P.J., Hanson, C.E., Eds.; Cambridge University Press: Cambridge, UK, 2007; pp. 717–743.

10. Füssel, H.-M. Adaptation planning for climate change: Concepts, assessment approaches, and key lessons. *Sustain. Sci.* **2007**, *2*, 265–275. [CrossRef]

11. Burton, I.; Huq, S.; Lim, B.; Pilifosova, O.; Schipper, E.L. From impacts assessment to adaptation priorities: The shaping of adaptation policy. *Clim. Policy* **2002**, *2*, 145–159. [CrossRef]

12. Burton, I. Do we have the adaptive capacity to develop and use the adaptive capacity to adapt? In *Climate Change, Adaptive Capacity and Development*; Smith, J., Klein, R., Huq, S., Eds.; Imperial College Press: London, UK, 2003; pp. 137–161.

13. Schipper, E.L.F. Conceptual history of adaptation in the UNFCCC process. *Rev. Eur. Community Int. Environ. Law* **2006**, *15*, 82–92. [CrossRef]

14. Huq, S.; Reid, H. Mainstreaming adaptation in development. *IDS Bull. Banner* **2004**, *35*, 15–21. [CrossRef]

15. Cassar, M.; Pender, R. The impact of climate change on cultural heritage: Evidence and response. In Proceedings of the ICOM Committee for Conservation: 14th Triennial Meeting, The Hague, The Netherlands, 12–16 September 2005; pp. 610–616.

16. Sabbioni, C.; Cassar, M.; Brimblecombe, P.; Lefevre, R.A. *Vulnerability of Clultural Heritage to Climate Change*; European and Mediterranean Major Hazards Agreement (EUR-OPA); Council of Europe: Strasburgo, France, 2008; pp. 1–24.

17. Heathcote, J.; Fluck, H.; Wiggins, M. Predicting and adapting to climate change: Challenges for the historic environment. *Hist. Environ. Policy Pract.* **2017**, *8*, 89–100. [CrossRef]

18. Haugen, A.; Mattsson, J. Preparations for climate change's influences on cultural heritage. *Int. J. Clim. Chang. Strateg. Manag.* **2011**, *3*, 386–401. [CrossRef]

19. Cassar, J. Climate change and archaeological sites: Adaptation strategies. In *Cultural Heritage from Pollution to Climate Change*; Lefèvre, R.-A., Sabbioni, C., Eds.; Edipuglia: Barri, Italy, 2016; pp. 119–127.

20. Pollard-Belsheim, A.; Storey, M.; Robinson, C.; Bell, T. The carra project: Developing tools to help managers identify and respond to coastal hazard impacts on archaeological resources. In Proceedings of the IEEE 2014 Oceans, St. John's, NL, Canada, 14–19 September 2014.

21. Carmichael, B.; Wilson, G.; Namarnyilk, I.; Nadji, S.; Brockwell, S.; Webb, B.; Hunter, F.; Bird, D. Local and indigenous management of climate change risks to archaeological sites. *Mitig. Adapt. Strateg. Glob. Chang.* **2018**, *23*, 231–255. [CrossRef]

22. Fatorić, S.; Seekamp, E. Evaluating a decision analytic approach to climate change adaptation of cultural resources along the atlantic coast of the united states. *Land Use Policy* **2017**, *68*, 254–263. [CrossRef]

23. Grøntoft, T. Climate change impact on building surfaces and façades. *Int. J. Clim. Chang. Strateg. Manag.* **2011**, *3*, 374–385. [CrossRef]

24. Hall, C.M.; Baird, T.; James, M.; Ram, Y. Climate change and cultural heritage: Conservation and heritage tourism in the anthropocene. *J. Herit. Tour.* **2015**, *11*, 10–24. [CrossRef]

25. Hall, C.M. Heritage, heritage tourism and climate change. *J. Herit. Tour.* **2015**, *11*, 1–9. [CrossRef]

26. Phillips, H. The capacity to adapt to climate change at heritage sites—The development of a conceptual framework. *Environ. Sci. Policy* **2015**, *47*, 118–125. [CrossRef]

27. Phillips, H. Adaptation to climate change at UK world heritage sites: Progress and challenges. *Hist. Environ. Policy Pract.* **2014**, *5*, 288–299. [CrossRef]

28. Fatorić, S.; Seekamp, E. Securing the future of cultural heritage by identifying barriers to and strategizing solutions for preservation under changing climate conditions. *Sustainability* **2017**, *9*, 2143. [CrossRef]

29. Carmichael, B.; Wilson, G.; Namarnyilk, I.; Nadji, S.; Cahill, J.; Bird, D. Testing the scoping phase of a bottom-up planning guide designed to support australian indigenous rangers manage the impacts of climate change on cultural heritage sites. *Local Environ.* **2017**, *22*, 1197–1216. [CrossRef]

30. Casey, A. Climate Change and Coastal Cultural Heritage: Insights from Three National Parks. Available online: https://digitalcommons.uri.edu/oa_diss/745/ (accessed on 9 August 2018).

31. Castellari, S.; Venturini, S.; Giordano, F.; Ballarin Denti, A.; Bigano, A.; Bindi, M.; Bosello, F.; Carrera, L.; Chiriacò, M.V.; Danovaro, R.; et al. *Elementi per una Strategia Nazionale di Adattamento ai Cambiamenti Climatici*; Ministero dell'Ambiente e della Tutela del Territorio e del Mare: Rome, Italy, 2014.

32. Castellari, S.; Venturini, S.; Pozzo, B.; Tellarini, G.; Giordano, F. *Analisi Della Normative Comunitaria e Nazionale Rilevante per gli Impatti, la Vulnerabilità e L'adattamento ai Cambiamenti Climatici*; Ministero dell'Ambiente e della Tutela del Territorio e del Mare: Roma, Italy, 2014.

33. Usmanov, B.; Nicu, I.C.; Gainullin, I.; Khomyakov, P. Monitoring and assessing the destruction of archaeological sites from Kuibyshev reservoir coastline, Tatarstan republic, Russian federation: A case study. *J. Coast. Conserv.* **2018**, *22*, 417–429. [CrossRef]

34. Nicu, I.C. Cultural heritage assessment and vulnerability using analytic hierarchy process and geographic information systems (Valea oii catchment, north-eastern Romania). An approach to historical maps. *Int. J. Disaster Risk Reduct.* **2016**, *20*, 103–111. [CrossRef]

35. Howard, A.J. Managing global heritage in the face of future climate change: The importance of understanding geological and geomorphological processes and hazards. *Int. J. Herit. Stud.* **2013**, *19*, 632–658. [CrossRef]

36. Cigna, F.; Tapete, D.; Lee, K. Geological hazards in the UNESCO world heritage sites of the UK: From the global to the local scale perspective. *Earth-Sci. Rev.* **2018**, *176*, 166–194. [CrossRef]

37. Smit, B.; Wandel, J. Adaptation, adaptive capacity and vulnerability. *Glob. Environ. Chang.* **2006**, *16*, 282–292. [CrossRef]

38. Lefèvre, R.-A. Resilience and adaptation of cultural heritage to climate change: International workshop in Ravello (Italy) 18–19 May 2017. In Proceedings of the 19th EGU General Assembly (EGU2017), Vienna, Austria, 23–28 April 2017; p. 3308.

39. Convention Concerning the Protection of the World Cultural and Natural Heritage. Available online: http://whc.unesco.org/archive/1995/whc-95-conf202-4reve.pdf (accessed on 14 August 2018).

40. Fatorić, S.; Seekamp, E. Are cultural heritage and resources threatened by climate change? A systematic literature review. *Clim. Chang.* **2017**, *142*, 227–254.

41. Michalski, D.J.; Bearman, C. Factors affecting the decision making of pilots who fly in outback Australia. *Saf. Sci.* **2014**, *68*, 288–293. [CrossRef]

42. Simonet, G.; Fatorić, S. Does "adaptation to climate change" mean resignation or opportunity? *Reg. Environ. Chang.* **2016**, *16*, 789–799. [CrossRef]

43. Poumadère, M.; Bertoldo, R.; Idier, D.; Mallet, C.; Oliveros, C.; Robin, M. Coastal vulnerabilities under the deliberation of stakeholders: The case of two french sandy beaches. *Ocean Coast. Manag.* **2015**, *105*, 166–176. [CrossRef]

44. Intergovernmental Panel on Climate Change. *Climate Change 2001: Impacts, Adaptation, and Vulnerability*; Cambridge University Press: Cambridge, UK, 2001.

45. Colette, A. *Climate Change and World Heritage. Report on Predicting and Managing the Impacts of Climate Change on World Heritage and Strategy to Assist States Parties to Implement Appropriate Management Responses*; UNESCO World Heritage Centre: Paris, France, 2007.

46. European Commission. *Adapting to Climate Change: Towards a European Framework for Action*; White Paper; Commission of the European Communities: Brussels, Belgium, 2009.

47. Della Torre, S. Oltre il restauro, oltre la manutenzione. In Proceedings of the International Conference Preventive and Planned Conservation, Monza, Mantova, 5–9 May 2014; pp. 1–10.

48. González Longo, C. Can Architectural Conservation Become Mainstream. Available online: https://strathprints.strath.ac.uk/50220/7/Gonzalez_Longo_ICOMOS_2015_Can_architectural_conservation_become_mainstream.pdf (accessed on 9 August 2018).

49. Marzeion, B.; Levermann, A. Loss of cultural world heritage and currenlty inhabited places to sea-level-rise. *Environ. Res. Lett.* **2014**, *9*, 1–7. [CrossRef]

50. Berenfeld, M.L. Climate change and cultural heritage: Local evidence, global responses. *George Wright Forum* **2008**, *25*, 66–82.

51. Intergovernmental Panel on Climate Change (IPCC). *Climate Change 2014: Impacts, Adaptation, and Vulnerability. Part b: Regional Aspects*; Contribution of Working Group II to the Fifth Assessment Report of the Intergovernmental Panel on Climate Change; Agard, J., Schipper, E.L.F., Birkmann, J., Campos, M., Dubeux, C., Nojiri, Y., Olsson, L., Osman-Elasha, B., Pelling, M., Prather, M.J., et al., Eds.; Cambridge University Press: New York, NY, USA, 2014; pp. 1757–1776.

52. Daire, M.-Y.; Lopez-Romero, E.; Proust, J.-N.; Regnauld, H.; Pian, S.; Shi, B. Coastal changes and cultural heritage (1): Assessment of the vulnerability of the coastal heritage in western france. *J. Isl. Coast. Archaeol.* **2012**, *7*, 168–182. [CrossRef]

53. A Climate Change Action Plan for Historic Scotland 2012–2017. Available online: https://www.historicenvironment.scot/media/2611/climate-change-plan-2012.pdf (accessed on 9 August 2018).

54. Curtis, R.; Hunnisett Snow, J. *Climate Change Adaptation for Traditional Buildings*; Historic Environment Scotland: Edinburgh, UK, 2016; pp. 1–55.

55. Directorate for Cultural Heritage. Climate Change and Cultural Heritage. Available online: https://edoc.site/ita-2014-world-tunnel-congress-3-pdf-free.html (accessed on 9 August 2018).

56. Anderson, J.; Artale, V.; Breil, M.; Gualdi, S.; Lionello, P. *From Global to Regional: Local Sea Level Rise Scenarios Focus on the Mediterranean Sea and the Adriatic Sea*; UNESCO: Venezia, Italy, 2010.

57. Mose Project Italy, Immersion of the Chioggia Flood Barrier Caisson. Available online: Availableonlinehttp://www.struktonimmersionprojects.com/news/2013/mose-project-italy---immersion-of-the--chioggia-flood-barrier-caissons/ (accessed on 9 August 2018).

58. Intergovernmental Panel on Climate Change (IPCC). *Climate Change 2007: Synthesis Report. Contribution of Working Groups i, ii and iii to the Fourth Assessment Report of the Intergovernmental Panel on Climate Change*; IPCC: Geneva, Switzerland, 2007; p. 104.

59. Agrawala, S. The European Alps: Location, economy and climate. In *Climate Change in the Alps: Adapting Winter Tourism and Natural Hazard Management*; Agrawala, S., Ed.; Organization for Economic and Co-Operation and Development: Paris, France, 2007.

60. Intergovernmental Panel on Climate Change (IPCC). *Climate Change 2014: Impacts, Adaptation, and Vulnerability. Part B: Regional Aspects*; Contribution of Working Group II to the Fifth Assessment Report of the Intergovernmental Panel on Climate Change; Agard, J., Schipper, E.L.F., Birkmann, J., Campos, M., Dubeux, C., Nojiri, Y., Olsson, L., Osman-Elasha, B., Pelling, M., Prather, M.J., et al., Eds.; Cambridge University Press: New York, NY, USA, 2014; pp. 117–130.

MDPI

St. Alban-Anlage 66

4052 Basel

Switzerland

Tel. +41 61 683 77 34

Fax +41 61 302 89 18

www.mdpi.com

Geosciences Editorial Office

E-mail: geosciences@mdpi.com

www.mdpi.com/journal/geosciences

www.ingramcontent.com/pod-product-compliance
Lightning Source LLC
Chambersburg PA
CBHW051856210326
41597CB00033B/5915